3D Printing in Medical Libraries

Medical Library Association Books

The Medical Library Association (MLA) features books that showcase the expertise of health sciences librarians for other librarians and professionals.

MLA Books are excellent resources for librarians in hospitals, medical research practice, and other settings. These volumes will provide health care professionals and patients with accurate information that can improve outcomes and save lives.

Each book in the series has been overseen editorially since conception by the Medical Library Association Books Panel, composed of MLA members with expertise spanning the breadth of health sciences librarianship.

Medical Library Association Books Panel

About the Medical Library Association

Founded in 1898, MLA is a 501(c)(3) nonprofit, educational organization of 3,500 individual and institutional members in the health sciences information field that provides lifelong educational opportunities, supports a knowledgebase of health information research, and works with a global network of partners to promote the importance of quality information for improved health to the health care community and the public.

Books in the Series

The Medical Library Association Guide to Providing Consumer and Patient Health Information, edited by Michele Spatz
Health Sciences Librarianship, edited by M. Sandra Wood
Curriculum-Based Library Instruction: From Cultivating Faculty Relationships to Assessment, edited by Amy Blevins and Megan Inman
The Small Library Manager's Handbook, by Alice Graves
Mobile Technologies for Every Library, by Ann Whitney Gleason
The Medical Library Association Guide to Answering Questions about the Affordable Care Act, edited by Emily Vardell

Marketing for Special and Academic Libraries: A Planning and Best Practices Sourcebook, by Patricia Higginbottom and Valerie Gordon
Interprofessional Education and Medical Libraries: Partnering for Success, edited by Mary E. Edwards
Translating Expertise: The Librarian's Role in Translational Research, edited by Marisa L. Conte
Expert Searching in the Google Age, by Terry Ann Jankowski
Digital Rights Management: The Librarian's Guide, edited by Catherine A. Lemmer and Carla P. Wale
The Medical Library Association Guide to Data Management for Librarians, edited by Lisa Federer
Developing Librarian Competencies for the Digital Age, edited by Jeffrey Coghill and Roger Russell
New Methods of Teaching and Learning in Libraries, by Ann Whitney Gleason
Becoming a Powerhouse Librarian: How to Get Things Done Right the First Time, by Jamie Gray
Assembling the Pieces of a Systematic Review: A Guide for Librarians, edited by Margaret J. Foster and Sarah T. Jewell
Information and Innovation: A Natural Combination for Health Sciences Libraries, edited by Jean P. Shipman and Barbara A. Ulmer
The Library Staff Development Handbook: How to Maximize Your Library's Most Important Resource, Mary Grace Flaherty
Transforming Medical Library Staff for the 21st Century, edited by Melanie J. Norton and Nathan Rupp
Health Sciences Collections Management for the Twenty-First Century, edited by Susan K. Kendall
The Medical Library Association Guide to Developing Consumer Health Collections, by Claire B. Joseph
Searching the Grey Literature: A Handbook for Finding Annual Reports, Working Papers, White Papers, Government Documents, and More, by Sarah Bonato
Transforming Health Sciences Library Spaces, edited by Alanna Campbell
3D Printing in Medical Libraries: A Crash Course in Supporting Innovation in Health Care, by Jennifer Herron

3D Printing in Medical Libraries

A Crash Course in Supporting Innovation in Health Care

Jennifer Herron

ROWMAN & LITTLEFIELD
Lanham • Boulder • New York • London

Published by Rowman & Littlefield
An imprint of The Rowman & Littlefield Publishing Group, Inc.
4501 Forbes Boulevard, Suite 200, Lanham, Maryland 20706
www.rowman.com

6 Tinworth Street, London SE11 5AL, United Kingdom

British Library Cataloguing in Publication Information Available

Library of Congress Cataloging-in-Publication Data

Names: Herron, Jennifer, 1987- author. | Medical Library Association, sponsoring body.
Title: 3D printing in medical libraries : a crash course in supporting innovation in health care / Jennifer Herron.
Description: Lanham : Rowman & Littlefield, [2019] | Series: Medical Library Association books series | Includes bibliographical references and index.
Identifiers: LCCN 2018041690 (print) | LCCN 2018042286 (ebook) | ISBN 9781538118801 (Electronic) | ISBN 9781538118795 (cloth : alk. paper) | ISBN 9781538125854 (pbk. : alk. paper)
Subjects: MESH: Printing, Three-Dimensional | Biomedical Technology | Libraries, Medical
Classification: LCC TS171.95 (ebook) | LCC TS171.95 (print) | NLM W 26.5 | DDC 621.9/8802461—dc23
LC record available at https://lccn.loc.gov/2018041690

∞ ™ The paper used in this publication meets the minimum requirements of American National Standard for Information Sciences Permanence of Paper for Printed Library Materials, ANSI/NISO Z39.48-1992.

Printed in the United States of America

Contents

Preface

Either write something worth reading or do something worth writing.
—Benjamin Franklin

I took Benjamin Franklin's advice a little too seriously, and back in November 2017, I submitted a book proposal about 3D printing in medical libraries. I thought I was doing something worth writing, and now, in 2018, I hope that I have written something worth reading.

I hope this book will serve as a guide to help start or further develop a 3D printing service in your medical or health sciences library or for any librarian trying to fulfill medical- and health-sciences-related requests. This book pulls from my experiences working at the Ruth Lilly Medical Library when I helped start the 3D printing service for students, faculty, and staff at Indiana University School of Medicine.

I had no prior experience with 3D printing or 3D modeling; the only skill I had was basic photo editing. Until the 3D printing service was being investigated, I had never really understood what 3D printing was and only assumed it was something more technical than I thought I could figure out—so technical, in fact, that I had had no interest in learning more about it and thought the complexity would be that of what it took to build robots. However, once the library showed interest in this service, my interest was piqued.

As I learned about 3D printing, I started to feel more confident and excited to try out this new technology. After taking the time to learn more about 3D printing, I was demystified, and it became a lot less intimidating. While some processes of 3D printing I still consider "muggle magic" (such as stereolithography [SLA], which uses ultraviolet light to create models), fuse deposition modeling (FDM), the method of 3D printing which involves extruding melted plastic, was instantly understandable to me when our library's tech team began visiting 3D print labs. Dr. Victor Soon, a faculty member, explained 3D printing using the FDM process in such a way that it suddenly all made sense—a "glorified hot-glue gun"—something low-tech to simplify for me these new high-tech 3D printers. It is my hope that I make 3D printing more approachable and take away the intimidation factor of learning a new technology like this, something outside the box of traditional librarianship skills.

I've arranged this book to gradually explore 3D printing and then to provide support for building and developing the service through marketing and outreach. I include information on finding models, editing models, and getting involved in the service itself. Troubleshooting was one of the most difficult challenges I encountered with 3D printing; whether it is my learning style or an inability to read an owner's manual, I struggled to understand the technical jargon associated with troubleshooting. I found YouTube to be the biggest support for this. Because there are a variety of learning styles, chapter 14 includes many materials and resources, varying from books to blogs to videos to social media channels.

Furthermore, I use a holistic approach to 3D printing; you can learn how to operate the printers, find the models, and troubleshoot problems, but if you do not understand what makes 3D printing—the desire to create and think outside the box—then you will struggle to market the service because you will not understand what message to send or who it should be sent to. What I'm referring to is the maker mind-set and design thinking. I've included resources to build your knowledge in this area and inspire and motivate the maker in you.

This book touches on all the elements that are important to learn for 3D printing overall; the applications to health sciences and medicine are woven throughout, with expert opinions featured in chapter 13 to provide outside viewpoints on 3D printing in a medical library.

I very much hope you enjoy this book and find it useful. If you have been on the fence about 3D printing, it is my goal that the information you find in this book will convince you that this is a worthwhile service. The information included in this book should provide you with guidance, resources, and support for launching, operating, and managing a 3D printing service or expand your knowledge if you have an established service.

Acknowledgments

First, I would have never made it through library school and written this book if not for my family. Their support has helped me through many things over the years and continued as they dropped off coffee and other forms of caffeine for me while I worked on this book.

To Gabe Rios and Kellie Kaneshiro: Without your approval and support, I would never have realized the true amazingness of 3D printing.

To the Ruth Lilly Medical Library staff: Thank you all for letting me run around and show you all the latest, unique requests that were printed and for supporting the 3D printing service by attending events and being promoters yourselves.

Last, thank you to all the librarians willing to explore this service area; I hope you see the potential and realize its impact on medical education and health-care innovation.

Chapter One

3D Printing

An Overview

Being able to print a physical model from a digital file seems like something from science fiction, but this technology is now widely available and accessible for everyday users. Soon, 3D printers may be as common as standard 2D inkjet printers and be found in every home, office, and business. Referred to as the third industrial revolution, social manufacturing (Suchow, 2016), and the individualized industrial revolution (Khoury, 2015), 3D printing has enabled everyday users to become designers and manufacturers and has fueled innovation across industries. While the name, 3D printing, is well known, there are a variety of terms that describe the process of creating a physical object from a computer design.

HISTORY OF 3D PRINTING

More technically speaking, 3D printing is the creation of a model, printed layer by layer, using a type of computer-aided design (CAD), stereolithography (STL), or object (OBJ) file as a map. *3D printing* is the more commonly used phrase to describe the consumer-use models; however, the phrases describing the process vary depending on the use and users of the technology. When describing industrial or commercial uses, *additive manufacturing* is frequently used. *Rapid prototyping* is the original phrase used to describe the production of 3D models for test products and manufacturing parts (Rayna and Striukova, 2016). Once 3D printing started to be used to create final products or pieces, *additive manufacturing* was more widely used by industrial users, with *3D printing* used by consumer users. *3D printing* also comes with its own variants and includes *3-dimensional printing*, *3-D printing*, and *3DP*, but with the rapid growth of consumer printers and their increasing affordability, *3D printing* is the more widely recognized phrase (Gebhardt, 2016).

Prior to 3D printing using the additive, layered approach, another technique used to create models was subtractive manufacturing. Subtractive manufacturing, also known as reductive manufacturing, operates just as the name implies: by removing material to leave only the model behind (Varotsis, n.d.a). The process resembles the art of woodworking and is completed using computer numerical control (CNC) machines for drilling, cutting, and grinding materials. The benefit of CNC machines is their ability to produce higher-resolution models with better accuracy than with the additive method. While CNC machines might produce models with higher accuracy, the main drawback for consumer use is the amount of wasted material and lack of cost savings. Starting with a solid block of material means that, to create the model, the unused material must be cut away, and, depending on the model, this could mean a lot of waste. It is estimated that metal CNC machines waste as much as 95 percent of material using the subtractive method. Additionally, for CNC machines to be cost-effective, they need to produce in mass quantities. 3D printing using the additive method prints only the model, with support material as needed. The amount of waste is much less: estimated to be near 5 percent (*Economist*, 2013).

3D PRINTING PROCESSES

Just as 3D-printing technology might go by different terms, the processes using this technology also go by different names. There are roughly seven methods used for 3D printing: fuse deposition modeling (FDM), stereolithography (SLA), digital light processing (DLP), selective laser sintering (SLS), material jetting, binder jetting, direct laser sintering (DMLS), and selective laser melting machine (SLMM) or electron beam melting (EBM) (3D Printing Industry, 2018; Varotsis, n.d.b). Of these seven methods, the most popular among consumer 3D printers is FDM and SLA.

FDM printers use a filament that is heated and then extruded through a nozzle onto a print bed, where it deposits filament, layer by layer, until the final model is completed. FDM printers can have a single nozzle and print in one material, or they can have multiple nozzles and print in a variety of materials and colors. Popular FDM printer brands include MakerBots, Ultimakers, LulzBot, and FlashForge. Additionally, 3D printer kits can be purchased, allowing for users to build their own printers. Some more mechanically inclined users have even built 3D printers using scrap computer parts.

SLA printers are more complex than FDM printers. SLA printers still create models layer by layer, but instead of building from the bottom up like FDM, they essentially build from the top down. SLA printers use a liquid material, resin, and an ultraviolet light to cure the resin, layer by layer. The consumer models of these printers usually only print in one material and color. Industrial printers have more capabilities and can print in multiple materials and colors. One of the most common SLA printer brands is Form-Labs. Higher-end SLA printers include Stratasys and 3D Systems. As with FDM printers, kits can be bought to self-assemble a printer, or instructions are available online for those users who wish to build from scratch.

3D PRINTING MATERIALS

Polylactic acid (PLA) plastic is known to be the easiest material to print with and good when starting out with 3D printing. Being easy to use is one of the major perks, along with the strength of the final model and its odorless printing. Furthermore, PLA can be sanded and painted for postprocessing touches. Two known drawbacks to PLA is the brittleness of the material and the difficulty gluing pieces together (3D Hubs, 2018; *3D Insider*, 2018; Rohringer, 2018; *Tinkercad Blog*, 2017).

Acrylonitrile butadiene styrene (ABS) plastic is another popular material used with FDM printers. It is stronger than PLA but can have some challenges when printing. ABS similarly can be painted and sanded, and furthermore, acetone can be used to join or weld pieces together. When used carefully, acetone can also smooth models and give a glazed-like appearance. Drawbacks to ABS involve an odor and potential issues with the fumes coming off the heated material.

Nylon is another material that is used for its strength and durability. A benefit of nylon is the lack of warping that can occur during printing when heated material encounters a cooler print bed. (More information on warping and its prevention can be found in chapter 9.) Like ABS, nylon also has an issue with fumes, and ventilation may be necessary.

Thermoplastic polyurethane (TPU), or thermoplastic elastomers (TPE), is a unique type of plastic that is flexible. TPU/TPE materials' flexibility is dependent on the design file. A more solid fill reduces the elasticity, whereas a hollow model has more flexibility. This flexibility does cause problems in postprocessing, as supports do not snap away as they do with other plastic materials.

Polyethylene terephthalate glycol (PETG) is a food-safe material that can be sterilized and is like a cross between PLA and ABS. The model has been used to create custom candy and food molds, as well as assorted packaging. PETG is commonly seen as the material used for water bottles. Another material that can print transparent models is polycarbonate (PC). PC is the strongest material in the plastic family and can provide extra protection to models as a clear covering. It is most commonly known for its use as lenses for eyeglasses.

With increased competition between 3D printer manufacturers, a need to go beyond plastic materials grew, and composite materials were created. Composite materials, sometimes considered exotic materials, are blends between PLA and ABS and other nonplastic materials. Examples of composites include wood, metal, and limestone. Because of some of these compositions, models can be magnetic and conductive, making them useful in creating devices (3D Hubs, 2018; *3D Insider*, 2018; Rohringer, 2018; *Tinkercad Blog*, 2017).

The numerous types of 3D printers on the market mean that material options are important to consider when reviewing 3D printers. Understanding the intended purpose of models and material available will help determine what type of 3D printer to invest in. The wide variety of material currently available and yet to come is believed to be what will continue to keep 3D printing thriving and persisting into the future.

FROM INDUSTRIAL TO CONSUMER USE

The shift to consumer 3D printing came about with the RepRap project. Wanting only to create a printer that could print its own parts, Dr. Adrian Bowyer, a Bath University professor, started a movement for at-home printing (Flynt, 2018). The Replicating Rapid Prototyper, or RepRap, was undertaken by Bowyer to determine if a 3D printer could, in a sense, self-replicate. The goal of the project was to make this low-cost printer available across the world and allow anyone access to the technology. RepRap was developed in 2004, but it gained popularity and collaborators in 2005 with publicity about Bowyer's efforts (Bensoussan, 2016). The RepRap project resulted in the creation of a series of printers, starting in 2007 with "Darwin," then "Mendel" and "Prusa Mendel" over the next few years, and ending in 2010 with "Huxley" (*All3DP*, 2016).

During the RepRap project, interest in 3D printing grew rapidly as printing services began to be offered, new printers came onto the market, and online resources for 3D-printable models emerged. Shapeways launched in 2008 and offered users the ability to print on demand by either using their library of models or by uploading their own designs. The following year, users had the ability to sell their products in Shapeways shops (Flynt, 2018). In 2009, MakerBot, a new build-it-yourself printer company, launched. The kit MakerBot provided enabled consumers to purchase, at a more affordable price, all the parts necessary to build their own 3D printers. To enable easier access to printable models, MakerBot created Thingiverse, an online resource for free 3D models for users to share, download, and print. Eventually, MakerBot began to sell their printers already built, but by this time, more 3D printer companies entered the consumer printer arena, and the market was flush with more affordable options for at-home 3D printing than ever before (Biggs, 2014). Kickstarter, an online organization that helps new businesses obtain funding, launched in 2009, allowing more and more 3D printing companies to find funding to get their businesses off the ground, and MakerBot now had to compete with 3D Systems and FormLabs, as well as a host of homegrown 3D printing start-ups (Bensoussan, 2016).

As interest in 3D printing has grown, so, too, has the technology, expanding to keep a competitive edge in the 3D printer market. MakerBot now offers a series of printers with features that range from extra-large print beds (Replicator Z-18) to new dual-material extruders and Smart+ extruders with enhanced performance measures for a higher-quality print. 3D Systems offers printers with dual- and triple-material extruders and an oversized printer for larger models.[1] As new upgrades were released, companies began competing to match each other's recent advancements. With so many printers on the market, to find an edge, materials became an important contributor to a printer's success or failure. Being able to print in multiple materials made some 3D printers more attractive, and when that became a standard option for printers, newer blended or composition materials were developed.

Industrial Use

Industries using 3D printing applications in manufacturing include aerospace, automotives, architecture, and consumer goods and electronics. The aerospace industry has taken great advantage of the new technology and has incorporated it into producing fuel nozzles and other jet engine parts. An advantage the aerospace industry found with 3D printing includes the ability to alter designs to create stronger, more efficient models. One example of this is the LEAP fuel nozzle. Engineers working on the 3D design adjusted the model, using the new capabilities of 3D printers, and increased the durability of LEAP nozzles five times over the original model (Zukas, 2015). Aerospace has pushed the limits of 3D printing further by using the technology in space. A specialized 3D printer that operates in zero gravity provides astronauts with the ability to print tools for repairs and even has printed a medical supply item: a finger splint (Strickland, 2017). NASA engineers feel a 3D printer also could provide buckles, clamps, containers, screws, caps, and springs (Committee on Space-Based Additive Manufacturing et al., 2014). Emergency repairs and the creation of specialized robots are other examples being investigated. The ability to reuse material and recycle models is another benefit and provides a sustainable resource for use in a limited environment.

The automotive industry similarly uses 3D printing for spare parts and design efficiencies (Desjardins, 2017). Ford uses 3D printing for a variety of car parts, such as the Ford ECOBoost engine. The ECOBoost engine received a custom intake manifold prior to the 2015 race 24 Hours of Daytona, and the new part helped the team win the race (3Ders, 2015). Ford continues to use 3D printing and is the first in the auto industry to use an industrial-sized 3D printer to experiment with custom auto parts and to alter existing part designs to make them lightweight in hopes of increasing vehicles' fuel economy (Ford Media Center, 2017). Outside of auto manufacturers, other arms of the auto industry have begun experimenting with 3D printing. Michelin created Vision, an airless tire, proving what the future holds for the auto industry (Low, 2017). The airless alternative is as durable as a standard tire, with the benefits of never needing inflation or new treads. Another unique aspect of the design is the material used; all materials to create the tire are biodegradable.

Consumer Use

Auto industries are not the only ones to benefit from 3D printing. Control nobs, custom cupholder covers, phone mounts, and other small replacement parts are examples of how consumers have taken advantage of 3D printing in the auto arena and beyond. By 2015, the availability of 3D printers rose to more than two hundred models providing multiple options. Consumer uses of 3D printers include a variety of knickknacks and action figures and figurines. Other uses of 3D printers revolve around fashion and tools. Beyond automotive replacement parts and pieces, consumers use 3D printing to fix broken parts of appliances or other household items (Zukas, 2015). 3D printing also helps support a new medium for artists and provides unique outlets for creativity. The multiple avenues available to share and sell these models allow a new type of art to emerge and enable a new means of funding for artists.

Societal Change

The multiple uses of 3D printing offer numerous benefits to industries and individuals; however, the impact on society is another aspect to be measured. The original method used in manufacturing, subtractive manufacturing, resulted in much waste of material. Because the real benefit using the subtractive method came from printing in mass quantities, this waste worsened with overproduction of models and products that did not sell (Suchow, 2016). Not only does the new additive method in 3D printing create less waste but also it allows for print on demand or as-needed models (Zukas, 2015). 3D printing also lessens waste by having the ability to print spare or replacement pieces for discontinued machinery or appliances. No longer do these items need to be thrown away; they now can be repaired and continue to be used. With the printers more accessible, they can operate within a factory or warehouse and eliminate the need for shipping, reducing waste even further.

Health and Medicine

In line with societal change, 3D printing is also making enormous changes in health and medicine and affecting patient care for the better. 3D printing in health and medicine is undergoing a massive surge of research. Health-care professionals from a variety of specialties and disciplines are exploring the use of 3D printing in medical education, as well as in clinical use and patient care, and the results are exciting. Even more exciting is learning about the many possibilities 3D printing offers to health care overall.

Medical Education

Health care can benefit in many ways from 3D printing, first with its impact on medical education. Faculty have taken advantage of 3D printing to enhance their instruction and have investigated various methods of implementing 3D printing into their curriculums and training for students and residents. One popular use is 3D-printed models of anatomy. In 2015, a study conducted at Macquarie University and Western Sydney University found that 3D-printed models of bones could be used in anatomy curriculums successfully. Collections of bones were scanned, made into digital models, and then printed for students to review. The researchers found that, when compared to the bones, the 3D-printed models had "no significant differences" (AbouHashem, Dayal, Savanah, and Strkalj, 2015). Additionally, bones previously too fragile to be examined by students could be scanned and duplicated, allowing them access to examine them.

In a study completed in 2014, researchers found 3D printing worked well filling gaps in limited resources in anatomy education. One frequent issue for anatomy education is the shortage of cadavers, which has led to an increase of virtual dissection-simulation systems; with these, though, students lose the physical, hands-on interaction available with cadavers. Furthermore, limiting students to visual displays can have a negative impact on tactile learners. Researchers at Monash University found that using 3D-printed models not only addressed the cadaver shortage but it also provided a safer alternative to the repeated exposure of students to cadavers containing embalming fluids with formalin. Faculty found the use of 3D printing advantageous to anatomy education for its speed, accuracy, and ability to print multiple copies of anatomical models to support multiple students in one lesson (McMenamin, Quayle, McHenry, and Adams, 2014).

While anatomy is one of the more obvious areas in student education where 3D printing has been successful, the use of 3D printing in creating simulations is also being reviewed. Maxillofacial surgeons have experimented using 3D-printed models based off patient scans to develop cases for students and residents to learn from and practice on and have found 3D printing to be "simple," "cost-efficient," and "promising" (Werz, Zeichner, Berg, Zeilhofer, and Thieringer, 2018). Further studies in this same line of investigation have all found 3D printing to be effective at replicating realistic surgical simulations and instruments (Lee et al., 2018) and have allowed for customized curriculums and enabled trainings for "removal of ocular foreign bodies, ultrasound-guided joint injections, nerve block injections, and various suturing and reconstruction procedures" (Lichtenberger et al., 2018).

While faculty have appreciated the flexibility and dependability of 3D printing to supplement their lessons, students also appreciate the additional support 3D models provide in their medical education. Students, especially those in their first years, have reported some weariness to cadaveric materials (Wisenden, 2018). When surveyed, first-year medical students at one medical school had mixed reactions to dissection, with some students excited and others finding the event "challenging" and "anxiety-provoking" (Hajj, 2015). Another study found student reactions to working with cadavers more positive, but results also identified an issue with students desiring more preparation before starting dissections (Cahill, 2009).

Multiple studies have compared 3D-printed models to cadaveric materials to determine whether 3D models have the potential to replace or supplement cadavers. Results from these studies indicate that 3D-printed models were not as realistic as cadaveric materials, but they ultimately found that students who used the 3D models more than the cadaveric materials perceived the models to be less intimidating to handle. Additionally, the customization that is possible with 3D printing and the ability to highlight and color-code specific anatomical sections supported student studying. The durability of models also gave students more freedom to examine models more thoroughly without fear of damaging the cadaveric material. Studies also found improved pre- and post-test scores for students who used 3D models (Lim, Loo, Goldie, Adams, and McMenamin, 2016; Mogali et al., 2018; Smith, Tollemache, Covill, and Johnston, 2018).

The examples of support that 3D printing provides to both faculty and students are numerous, and results have shown 3D printing to be a new resource to support health and medical education. Furthermore, the lower costs involved in printing a model versus buying a professionally made model also make 3D printing lucrative. The exposure to 3D-printing technology also engages a higher level of thinking for some students who are interested in learning how to apply the technology to patient care. Providing experiences with this technology while still learning about health care also has the

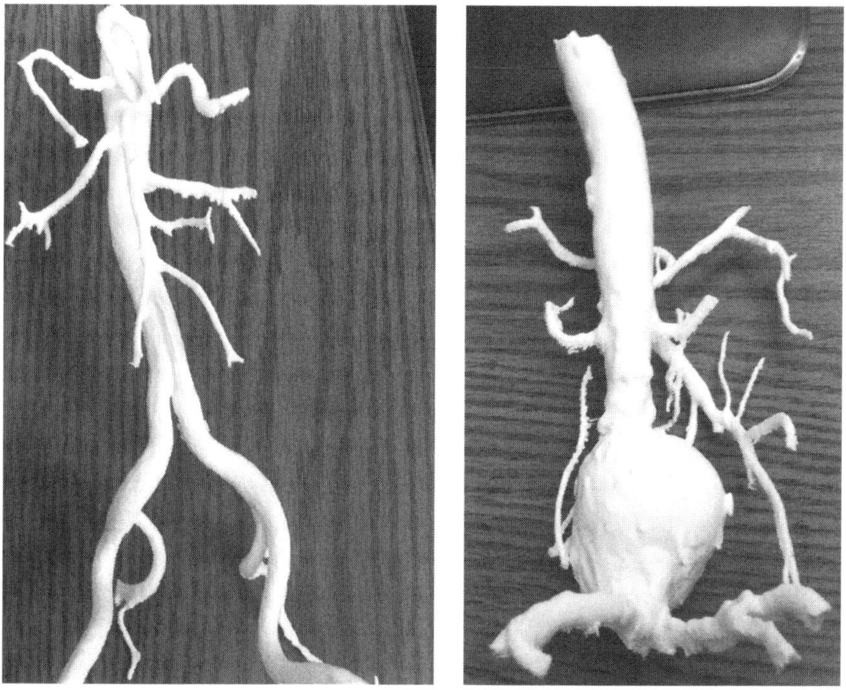

Figure 1.1. Using 3D printing for anatomy education. Printing as one piece or multiple pieces gives students an in-depth look at anatomy. *Jennifer Herron.*

potential to gain fresh, new ideas from students and provides faculty with ideas to spur new research and applications.

Clinical Care

Beyond helping future health-care professionals, 3D printing is also helping those currently in practice. Using 3D-printed models for anatomical purposes as discussed earlier has proven to be a useful resource and easily transitions from anatomy education to presurgical planning. Numerous studies have reported benefits for presurgical planning, including increased precision, decreased operating time, cost-efficiency, and improved outcomes, and often describe the 3D-printed models as being "invaluable" (Dickinson et al., 2015; Farooqi, Gonzalez-Lengua, Shenoy, Sanz, and Nguyen, 2016; Hughes et al., 2017; Mukherjee, Cheng, Flanagan, and Greenberg, 2017; Pacione, Tanweer, Berman, and Harter, 2016; Yoo and van Arsdell, 2017).

In addition to aiding health-care professionals plan treatments, 3D-printed models have also been shown to aid patients in understanding their treatment and conditions. Multiple studies have found that reviewing 3D-printed models of their medical scans increases patients' knowledge of their conditions, makes for more effective appointments, facilitates better acceptance of their conditions, and makes them more comfortable with upcoming procedures. Additionally, 3D-printing patient-specific models and using them in the explanation of their conditions and treatment plans gave some patients a sense of control of their health (Biglino et al., 2015; Biglino et al., 2017; Eisenmenger, Wiggins, Fults, and Huo, 2017; Yang et al., 2018).

3D printing is unique in that patients also can get involved and use the technology in their health care. One patient proactively used 3D printing for health care in 2013. Michael Balzer's wife had been diagnosed with a brain tumor behind her eye, and it was recommended that she have it surgically removed. Balzer, experienced with 3D graphics, requested the MRI scans and was able to create a 3D model of his wife's tumor. After a second scan showed an increase in tumor size, Balzer created another model and compared the results. Incredibly, Balzer found that a diagnosis error had been made, and his wife's tumor in the second scan had been measured incorrectly. Despite the error, Balzer's wife still needed her tumor removed, which prompted him to 3D-print a model. Using both his 3D virtual model and his 3D-printed model, Balzer consulted experts across the world for opinions on a less-invasive surgery that would reduce risk. As a result of his efforts, Balzer found a new type of surgical procedure to investigate with his wife's surgeons and changed the course of her care (Breselor, 2015).

Michael Balzer and his wife, Pamela Shavaun Scott, are a unique example of how patients can use 3D printing for their health care. While not all patients will have the same technical skills that Balzer has, the same resources are available, and patients' knowledge of 3D printing's potential may soon grow to the point where they request 3D prints. Engaging patients to become more proactive has been a movement in the health-care world for some time, and with 3D printing's ability to increase patient understanding of conditions and treatments, it might be a step in the right direction for increasing patient engagement.

Low-Cost Supplies and Devices

The direct benefits of 3D printing for students, faculty, health-care professionals, and patients can be found in numerous studies. 3D printing has made great advances in filling gaps in health care in developing countries. The lower costs associated with models from a 3D printer also have given way to the idea of 3D printers acting as the "doctor's bag of the future" (Strickland, 2017). The ability to print medical instruments and devices is being investigated by numerous agencies and organizations. A pair of New Zealand doctors, Dr. Hong Sheng Chiong and Dr. Benjamin O'Keeffe, formed oDocs and have created an open-source, printable slit-lamp attachment and fundus camera that, once condensing lenses are added, works with an iPhone to become a portable eye-care kit (Clarke, 2017). They hope that the decreased cost and open-source platform will enable health-care providers all over the world to provide eye exams to those in need.

Field Ready, a humanitarian relief group that partners technology experts with aid workers, has also taken advantage of 3D printing to help build medical supplies for clinics in developing countries. After the 2015 earthquake in Nepal, a local health clinic struggled to replace supplies after paying for repairs. Health-care workers in Bhotechaur struggled to treat patients due to a lack of supplies and equipment, but a member of Field Ready was able to provide some help by designing a 3D-printable otoscope (3Ders, 2017). Field Ready has also previously assisted in medical issues by 3D-printing parts for malfunctioning and broken medical devices and printing some basic medical supplies, such as ECG clamps and otoscope specula (Field Ready, 2016).

In January 2017, astronaut and doctor Julielynn Wong used a 3D printer in space to print a finger splint that was customized and fitted using previous scans of her hand. 3D-printing medical supplies was not something new for Dr. Wong; however, the ability to do so in such an environment shined a light on just what capabilities 3D printers can have in health care. Years prior to her trip to space, Dr. Wong established 3D4MD, an organization dedicated to using 3D printers, to help address health-care disparities in low-income and developing countries, as well as for those in remote and isolated areas (3D4MD, 2018). Other initiatives like Dr. Wong's follow along the lines of the maker movement, where problems are solved by making items, usually by altering existing items or by creating brand-new items to solve a problem at hand. To better prepare students and professionals for this new age in health and medicine, design thinking has emerged and is slowly being integrated into some school curriculums. While encouraging students and professionals to think outside the box, it is also challenging, as the law is trying to keep up with technology and the overwhelming amount of change.

NOTE

1. Oversized printer (Cube Pro Trio) is no longer available for consumer printing.

REFERENCES

3D4MD. 2018. "About." http://www.3d4md.com/about.

3Ders. 2015. "Ford Uses 3D Printed Parts in ECOboost Race Engine . . . and Wins 24 Hours of Daytona Race." https://www.3ders.org/articles/20150530-ford-uses-3d-printed-parts-in-ecoboost-race-engine-and-wins-24-hours-of-daytona-race.html.

———. 2017. "3D Printing an Increasing Source of Medical Aid in Earthquake-Devastated Nepal." http://www.3ders.org/articles/20170306-3d-printing-an-increasing-source-of-medical-aid-in-earthquake-devastated-nepal.html.

3D Hubs. 2018. "FDM 3D Printing Materials Compared." https://www.3dhubs.com/knowledge-base/fdm-3d-printing-materials-compared.

3D Insider. 2018. "16 Types of 3D Printer Filaments." http://3dinsider.com/3d-printing-materials.

3D Printing Industry. 2018. "The Free Beginner's Guide: 3D Printing Industry." https://3dprintingindustry.com/3d-printing-basics-free-beginners-guide.

AbouHashem, Y., M. Dayal, S. Savanah, and G. Strkalj. 2015. "The Application of 3D Printing in Anatomy Education." *Medical Education Online* 20, no. 1: 29847. DOI: 10.3402/meo.v20.29847.

All3DP. 2016. "The Official History of the RepRap Project." https://all3dp.com/history-of-the-reprap-project.

Bensoussan, H. 2016. "The History of 3D Printing: 3D Printing from the 80s to Today." Sculpteo. December 14, 2016. https://www.sculpteo.com/blog/2016/12/14/the-history-of-3d-printing-3d-printing-technologies-from-the-80s-to-today.

Biggs, J. 2014. "The Promise and Perils of 3D Printer Popularity." TechCrunch. https://techcrunch.com/2014/08/28/the-promise-and-perils-of-3d-printer-popularity.

Biglino, G., C. Capelli, L.-K. Leaver, S. Schievano, A. M. Taylor, and J. Wray. 2015. "Involving Patients, Families and Medical Staff in the Evaluation of 3D Printing Models of Congenital Heart Disease." *Communication and Medicine* 12, nos. 2–3: 157–69.

Biglino, G., D. Koniordou, M. Gasparini, C. Capelli, L.-K. Leaver, S. Khambadkone, S. Schievano, A. M. Taylor, and J. Wray. 2017. "Piloting the Use of Patient-Specific Cardiac Models as a Novel Tool to Facilitate Communication during Cinical Consultations." *Pediatric Cardiology* 38, no. 4: 813–18. DOI: 10.1007/s00246-017-1586-9.

Breselor, S. 2015. "Man Saves Wife's Sight by 3D Printing Her Tumor." *Make:*. January 14, 2015. https://makezine.com/2015/01/14/hands-on-health-care.

Cahill, K. C., and R. R. Ettarh. 2009. "Attitudes to Anatomy Dissection in an Irish Medical School." *Clinical Anatomy* 22, no. 3: 386–91. DOI: 10.1002/ca.20777

Clarke, C. 2017. "oDocs Launches Open Source 3D Printed Eye Care Kit." 3D Printing Industry. May 30, 2017. https://3dprintingindustry.com/news/odocs-launches-open-source-3d-printed-smartphone-eye-care-kit-114549.

Committee on Space-Based Additive Manufacturing, Aeronautics and Space Engineering Board, National Materials and Manufacturing Board, Division on Engineering and Physical Sciences, and National Research Council. 2014. *3D Printing in Space*. Washington, DC: National Academies Press.

Desjardins, J. 2017. "Infographic: All the Ways 3D Printing Is Changing the World." *Business Insider*. http://www.businessinsider.com/infographic-3d-printing-2017-9.

Dickinson, K. J., J. Matsumoto, S. D. Cassivi, J. M. Reinersman, J. G. Fletcher, J. Morris, L. M. Wong Kee Song, and S. H. Blackmon. 2015. "Individualizing Management of Complex Esophageal Pathology Using Three-Dimensional Printed Models." *Annals of Thoracic Surgery* 100, no. 2: 692–97. DOI: 10.1016/j.athoracsur.2015.03.115.

Economist. 2013. "Adding and Taking Away: A 3D Printer Meets a Milling Machine." *Babbage* (blog). https://www.economist.com/blogs/babbage/2013/12/advanced-manufacturing.

Eisenmenger, L. B., R. H. Wiggins III, D. W. Fults III, and E. J. Huo. 2017. "Application of 3-Dimensional Printing in a Case of Osteogenesis Imperfecta for Patient Education, Anatomic Understanding, Preoperative Planning, and Intraoperative Evaluation." *World Neurosurgery* 107: 1049.e1–7. DOI: 10.1016/j.wneu.2017.08.026.

Farooqi, K. M., C. Gonzalez-Lengua, R. Shenoy, J. Sanz, and K. Nguyen. 2016. "Use of a Three Dimensional Printed Cardiac Model to Assess Suitability for Biventricular Repair." *World Journal of Pediatric and Congenital Heart Surgery* 7, no. 3: 414–16. DOI: 10.1177/2150135115610285.

Field Ready. 2016. "3D Printing Spare Parts for Vital Medical Equipment in the Field." https://www.fieldready.org/single-post/2016/10/14/3D-Printing-Spare-Parts-for-Vital-Medical-Equipment-in-the-Field.

Flynt, J. 2018. "History of 3D Printing Timeline: Who Invented 3D Printing?" *3D Insider*. http://3dinsider.com/3d-printing-history.

Ford Media Center. 2017. "Ford Tests Large-Scale 3D Printing with Light-Weighting and Personalization in Mind." https://media.ford.com/content/fordmedia/fna/us/en/news/2017/03/06/ford-tests-large-scale-3d-printing.html.

Formlabs. n.d. "The Ultimate Guide to Stereolithography (SLA) 3D Printing." Accessed May 8, 2018. https://formlabs.com/blog/ultimate-guide-to-stereolithography-sla-3d-printing.

Gebhardt, A., and J.-S. Hötter. 2016. *Additive Manufacturing: 3D Printing for Prototyping and Manufacturing*. Munich: Carl Hanser Verlag.

Hajj, I., M. Dany, W. Forbes, et al. 2015. "Perceptions of Human Cadaver Dissection by Medical Students: A Highly Valued Experience." *Italian Journal of Anatomy and Embryology* 120, no. 3: 162–71. https://www.ncbi.nlm.nih.gov/pubmed/27086415.

Hughes, A. J., C. DeBuitleir, P. Soden, B. O'Donnchadha, A. Tansey, A. Abdulkarim, C. McMahon, and C. J. Hurson. 2017. "3D Printing Aids Acetabular Reconstruction in Complex Revision Hip Arthroplasty." *Advances in Orthopedics* 2017: 8925050. DOI: 10.1155/2017/8925050.

Khoury, A. H. 2015. "The Makings of an 'Individualized-Industrial' Revolution: Three-Dimensional Printing and Its Implications on Intellectual Property Law." *Journal of High Technology Law* 16, no. 1. https://ssrn.com/abstract=2964567.

Lee, S., J. Y. Ahn, M. Han, G. H. Lee, H. K. Na, K. W. Jung, J. H. Lee, D. H. Kim, K. D. Choi, H. J. Song, and H. Y. Jung. 2018. "Efficacy of a Three-Dimensional-Printed Training Simulator for Endoscopic Biopsy in the Stomach." *Gut and Liver* 12, no. 2: 149–57. DOI: 10.5009/gnl17126.

Lichtenberger, J. P., III, P. S. Tatum, S. Gada, M. Wyn, V. B. Ho, and P. Liacouras. 2018. "Using 3D Printing (Additive Manufacturing) to Produce Low-Cost Simulation Models for Medical Training." *Military Medicine* 183 (suppl 1): 73–77. DOI: 10.1093/milmed/usx142.

Lim, K. H., Z. Y. Loo, S. J. Goldie, J. W. Adams, and P. G. McMenamin. 2016. "Use of 3D Printed Models in Medical Education: A Randomized Control Trial Comparing 3D Prints versus Cadaveric Materials for Learning External Cardiac Anatomy." *Anatomical Sciences Education* 9, no. 3: 213–21. DOI: 10.1002/ase.1573.

Low, C. 2017. "Michelin's 3D-Printed Tire Is as Stunning as It Is Futuristic." Engadget. August 7, 2017. https://www.engadget.com/2017/08/07/michelin-vision-biodegradable-3d-print-airless-tire.

McMenamin, P. G., M. R. Quayle, C. R. McHenry, and J. W. Adams. 2014. "The Production of Anatomical Teaching Resources Using Three-Dimensional (3D) Printing Technology." *Anatomical Sciences Education* 7, no. 6: 479–86. DOI: 10.1002/ase.1475.

Mogali, S. R., W. Y. Yeong, H. K. J. Tan, G. J. S. Tan, P. H. Abrahams, N. Zary, N. Low-Beer, and M. A. Ferenczi. 2018. "Evaluation by Medical Students of the Educational Value of Multi-Material and Multi-Colored Three-Dimensional Printed Models of the Upper Limb for Anatomical Education." *Anatomical Sciences Education* 11, no. 1: 54–64. DOI: 10.1002/ase.1703.

Mukherjee, P., K. Cheng, S. Flanagan, and S. Greenberg. 2017. "Utility of 3D Printed Temporal Bones in Pre-surgical Planning for Complex BoneBridge Cases." *European Archives of Oto-Rhino-Laryngology* 274, no. 8: 3021–28. DOI: 10.1007/s00405-017-4618-4.

Pacione, D., O. Tanweer, P. Berman, and D. H. Harter. 2016. "The Utility of a Multimaterial 3D Printed Model for Surgical Planning of Complex Deformity of the Skull Base and Craniovertebral Junction." *Journal of Neurosurgery* 125, no. 5: 1194–97. DOI: 10.3171/2015.12.jns151936.

Rayna, T., and L. Striukova. 2016. "From Rapid Prototyping to Home Fabrication: How 3D Printing Is Changing Business Model Innovation." *Technological Forecasting and Social Change* 102: 214–24. https://doi.org/10.1016/j.techfore.2015.07.023.

Rohringer, S. 2018. "2018 3D Printer Filament Guide—All You Need to Know." *All3DP*. https://all3dp.com/1/3d-printer-filament-types-3d-printing-3d-filament.

Smith, C. F., N. Tollemache, D. Covill, and M. Johnston. 2018. "Take Away Body Parts! An Investigation into the Use of 3D-Printed Anatomical Models in Undergraduate Anatomy Education." *Anatomical Sciences Education* 11, no. 1: 44–53. DOI: 10.1002/ase.1718.

Strickland, A. 2017. "The 'Doctor's Bag of the Future' Could Be a 3-D Printer." CNN. https://www.cnn.com/2017/08/15/health/medical-3d-printing-body-smart/index.html.

Suchow, R. 2016. "3D Printing and Service Learning: Social Manufacturing as a Vehicle for Developing Social Awareness." *Journal of Catholic Education* 20, no. 1: 292–300. DOI: 10.15365-joce.2001132016.

Tinkercad Blog. 2017. "3D Printing Materials Guide." June 27, 2012. https://blog.tinkercad.com/materialsguide.

Varotsis, A. B. n.d.a. "3D Printing vs. CNC Machining." 3D Hubs. Accessed May 8, 2018. https://www.3dhubs.com/knowledge-base/3d-printing-vs-cnc-machining.

———. n.d.b. "Selecting the Right 3D Printing Process." 3D Hubs. Accessed May 8, 2018. https://www.3dhubs.com/knowledge-base/selecting-right-3d-printing-process.

Werz, S. M., S. J. Zeichner, B. I. Berg, H. F. Zeilhofer, and F. Thieringer. 2018. "3D Printed Surgical Simulation Models as Educational Tool by Maxillofacial Surgeons." *European Journal of Dental Education* 22, no. 3: e500–e505. DOI: 10.1111/eje.12332.

Wisenden, P. A. 2018. "Emotional Response of Undergraduates to Cadaver Dissection." *Clinical Anatomy* 31, no. 2: 224–40. DOI: 10.1002/ca.22992

Yang, T., T. Tan, J. Yang, J. Pan, C. Hu, J. Li, and Y. Zou. 2018. "The Impact of Using Three-Dimensional Printed Liver Models for Patient Education." *Journal of International Medical Research* 46, no. 4: 1570–78. DOI: 10.1177/0300060518755267.

Yoo, S.-J., and G. S. van Arsdell. 2017. "3D Printing in Surgical Management of Double Outlet Right Ventricle." *Front Pediatrics* 5: 289. DOI: 10.3389/fped.2017.00289.

Zukas, V., and J. A. Zukas. 2015. "An Introduction to 3D Printing." http://search.ebscohost.com/login.aspx?direct=true&db=e700xna&AN=1003128&site=ehost-live.

Chapter Two

Legal Concerns Involving 3D Printing

3D printing is often referred to as a disruptive technology with regards to its revolutionary abilities; however, in the legal realm, it continues this disruptive trend. 3D printing in general can cause issues with intellectual property (IP), trademarks, patents, and copyright. Beyond infringing on these laws, liability becomes a concern, as well as other potential health-care impacts.

INTELLECTUAL PROPERTY

Intellectual property rights and ownership has been a much-contested issue in recent years due to the increased technologies that allow for easier-than-ever sharing of content online. The law has struggled with newer uses of technology, trying to determine where rights lie with regards to digital technology, as was the case with, Naruto, a San Francisco monkey whose published selfie started a debate on intellectual property ownership and whether animals in fact have ownership rights to pictures they take (Cullinane, 2018). 3D printing adds another level of complexity between technology and the law when dealing with intellectual property rights. 3D printing uses the designs from the minds of artists, engineers, and anyone well versed with computer-aided design, or the creation a 2D of 3D model using a computer.

Intellectual property rights provide protections for IP owners with copyright, trademark, and patent rights. These are different mechanisms by which IP owners are recognized for their work and protected from others who may infringe on their ownership. 3D printing threatens not only big commercial industries but individuals as well in all these regards.

COPYRIGHT, PATENTS, TRADEMARKS

Copyright allows for IP owners to profit off their work and prevent others from profiting or unfairly reusing or changing their work. Librarians will have some familiarity with copyright, as it pertains to the sharing of articles from paid, subscription journals (World Intellectual Property Organization, 2011). Virtual models are protected, just as digital music is protected. Both the virtual 3D model and printed model can be protected by copyright and therefore cannot be reproduced or altered without permission (Chan, 2015). Within copyright, there are some exceptions to the law. If a model is deemed useful and not a distinct and creative work, then copyright does not apply. A button could be something considered useful, but if the button is ornate or includes a decorative element, then it edges away from useful and more toward a creative work.

A patent is another way for IP owners to share their work while still retaining control over its use. Patents allow for the IP owner to give permissions to others to reproduce their work, but this is only regarding inventions or new processes (World Intellectual Property Organization, 2011). Patents may become more of an issue in 3D printing as devices are built from printed parts. The method of printing these parts (speed, temperature, resolution) may also be subject to patentability. 3D model files now also fall under patentability due to the lack of copyright protections available for the digital files. With a patent, the 3D model itself receives protection and, with it, the digital file to the crucial element that, without it, no model could be printed. By becoming a crucial factor of the final product, the file gains patent protections and cannot be shared without permission from the owner (Malaty and Rostama, 2017).

Patent violations are becoming more frequent as new, homegrown printers and even higher-end printers are not distinct enough from similar products in the marketplace. An example of this patent infringement can be seen in the case of EnvisionTEC versus FormLabs, where EnvisionTech argues the SLA process used by FormLabs is too similar to their printers (Hornick and Wells, 2016).

The abundance of similar products in the market created a need for trademarks, and IP owners can register for a trademark to distinguish their work from others. Trademarks generally imply quality and that the standards of the product meet those set by the trademark owner. Fake reproductions sometimes use trademarks to fool consumers into paying more money than an item is worth. The ease of scanning and copying items and reproducing them with a printer is one reason trademarks become an issue: They make it more challenging to identify the fake products when the replications are of good quality. John Hornick (2015) aptly describes the issue, stating that 3D printers may act as a "steroid for counterfeiters" (813). In a 2016 *3D Printing*

Industry article, Nick Hall wrote that, in 2018, trademark violations will cost businesses more than $100 billion, showing just how big of a problem trademark owners are facing.

Trademark violation is the aspect of IP where consumers who print trademarked 3D items can run into legal trouble. When 3D printers became more popular, new businesses formed and offered 3D models for purchase to print. These models generally include a license agreement and stipulations for its use. Sharing these model files violates the license agreement, and like the legal problems file-sharing companies like Napster faced, companies like Thingiverse may fall into trouble if user uploads break these agreements.

There have not been any major cases in the news regarding private 3D printing infringing on major businesses or companies, which may be because there is not a good system in place to alert IP holders or those who enforce IP laws about violations. By the time the violations are caught, duplicate files most likely exist elsewhere on the internet, where continued violations can occur until it is discovered. Furthermore, tracking reproductions made using at-home printers is practically impossible unless items are sold.

Another growing concern with 3D printing involves criminal activity beyond IP violations. The first 3D-printed gun was tested in 2013. Dubbed the Liberator and developed by Defense Distributed, it was printed using ABS plastic (Morelle, 2013). As 3D printing becomes more prevalent, the dangers also become apparent. 3D-printed weapons can be virtually untraceable. Beyond printed guns, other dangerous models are bomb parts. While less threatening, additional models for criminal purposes include parts for credit-card skimmers. In 2014, a group in Italy, France, Spain, and Germany had printed more than one hundred devices before they were arrested (Chase and LaPorte, 2018).

LIABILITY

Liability is one area where health and medical 3D printing differ from general 3D printing. Liability involves holding someone responsible for damage or harm, and 3D-printed medical and health-related models can fall into trouble with liability in a few ways. The previous chapter highlights the varying uses of 3D printing in health and medicine, and while these uses have been researched and tested, safety can become an issue due to defective prints or unforeseen consequences. One easily correctable way to eliminate liability with 3D printing, for both health-related and general models, is to ensure that postprocessing has been completed. 3D-printed models that are not properly postprocessed may have sharp edges and could harm someone if not sanded or smoothed. This is one example of a basic level of liability; however, once a model is used in the treatment of a patient, it enters a new realm of liability: medical liability.

Medical liability occurs when a health-care provider causes harm to a patient through a negligent act. One simple way this can occur is if a model given to a patient has a sharp edge. Another way is if a health-care professional provides a 3D-printed prosthetic that causes injury due to improper fit. An abundance of prosthetic models are freely available online, and a good-intentioned student or other health-care professional may inadvertently cause harm if not careful.

When 3D printing is used during surgery, not only do the printed models become liabilities but the virtual models, as well. One example is when software used to create a patient-specific 3D model in turn generates a patient-specific printed model. A 3D model used as a cutting guide required a virtual model that was just as important to the end product, so both the printed model and the virtual model could be held liable in court (Beck and Jacobson, 2017).

HIPAA

The Health Insurance Portability and Accountability Act (HIPAA) protects patients from having their medical information released without their permission. The protected health information (PHI) is any type of information that could result in a patient being identified. Typically, PHI includes names, addresses, birth dates, and social security numbers; however, with 3D printing, there is a new means to identify patients (US Department of Health and Human Services, 2013). Chapter 7 highlights several websites that involve patient-specific models. It is the responsibility of health-care professionals to ensure that HIPAA regulations have been followed in the creation of 3D models. If they decide to share a model to any of these sites, they must also ensure that they do not inadvertently reveal any patient information in the process. PHI may be inadvertently released with a file name or a model that includes identifiable features. With all the potential HIPAA violations possible with 3D printing, the real problem lies with sharing model files. While most of these files are shared for educational purposes, some are shared by organizations that do not have access to a 3D printer. By printing in-house, potential HIPAA violations are reduced, which is one benefit of having 3D print labs on site (Naftulin, Kimchi, and Cash, 2015).

FDA INVOLVEMENT

In their efforts to advance patient care, health-care professionals must also follow Food and Drug Administration (FDA) rules and regulations with the safety of their models. Models printed for surgical planning must be created with FDA-approved software and printed with FDA-approved material and with an FDA-approved printer. Just as certain displays are FDA approved for reviewing MRI and CT scans, the software used to generate the 3D model must also follow the standards laid out by the FDA. Depending on the intended use of models, certain materials are also required to be handled and stored in specific ways in order to maintain FDA approval.

In late 2017, the FDA released a draft guidance to help people who are 3D-printing devices for medical use. In a press release, FDA Commissioner Scott Gottlieb called it a "leap-frog guidance" to "bridge where we are today with innovations of tomorrow" (US Food and Drug Administration, 2017b).

The guidance was just that—a series of recommendations to help provide insight into the processes used by the FDA to approve 3D-printed medical devices. The draft guidance provides information on the design and manufacturing review, as well as the testing process manufacturers should review as they work on developing their devices (US Food and Drug Administration, 2017c). In addition to providing background information on the device-approval process, the FDA also released information on device classifications. The FDA sorts medical devices into three categories: class I, general controls; class II, general controls and special controls; and class III, general controls and premarket approval. Which class devices fall into depends on the intended use and indications of use, and as such, the market-submission requirements for each class get more intensive (US Food and Drug Administration, 2018).

The FDA also provides a sample workflow, identifying the possible steps taken when 3D-printing a medical device and what should be considered during each step to help obtain approval from the FDA. The following are the steps and what occurs, or should be considered, during each step:

- Device design (3D model of medical device)
- Software workflow (3D model print file)
- Materials controls (quality and standards of the materials used in printing)
- Printing (actual printing process)
- Postprocessing (removal of support, sanding, sterilization, etc.)
- Process validation and verification (checking printed materials for accuracy and usability)
- Testing (testing to confirm the device is safe for medical use and performs as stipulated by its intended use; US Food and Drug Administration, 2017a)

SUGGESTIONS

The following are some suggestions to avoid any potential issues with IP. I also recommend these suggestions as a best practice for creating and developing an open line of communication with your organization's legal department.

Make Contact!

Don't wait to be contacted about 3D printing. It will most likely be on better terms if you contact your organization's legal department than if they need to contact you. Make your 3D-printing-service operation known to the legal department to enable a point of contact between them and your service. Explain your service and what you have been printing and allow them to gain a better understanding of what you are doing instead of later finding out about a problem.

Know What You Can and Cannot Print

Having a policy that explains your service and what is and is not allowed to be printed not only helps you with questionable print requests but also ensures administrators that you are aware of potential issues and are already taking action.

Create a Disclaimer, Agreement, or Acceptable-Use Form

Public computers often include an acceptable-use policy when logging on, and 3D printers should not be treated differently. You cannot predict whether a good intention might go badly, and you want to take responsibility off not only yourself but the library as well. Similarly, it is almost impossible to determine the source of some models. Unless you find model files yourself to fill requests or receive a self-designed file, you are receiving files to print and are not able to track the owner. If requestors ask for a file to be printed or if you want a self-serve lab, then you need to have some method of safeguarding against copyright infringements and trademark violations. If you contact your organization's legal department, they might be able to assist you with the wording for request forms to prevent problems from coming back on the 3D printing service.

Make a Flowchart for Innovation and Where the Library's 3D Printing Service Falls

Clearly identify where on the health-care-innovation spectrum the library's 3D printing service falls. For the areas where the library needs to step back, see if you can find alternative resources for referral and list them on the flowchart. Again, administrators will appreciate seeing how the library is involved.

REFERENCES

Beck, J. M., and M. D. Jacobson. 2017. "3D Printing: What Could Happen to Products Liability When Users (and Everyone Else in Between) Become Manufacturers." *Minnesota Journal of Law, Science, and Technology* 18, no. 1: 143. https://scholarship.law.umn.edu/mjlst/vol18/iss1/3.

Chan, J. R. M. 2015. "Copyright Considerations for Providing 3D Printing Services in the Library." *Bulletin of the Association for Information Science and Technology* 42: 26–31. DOI:10.1002/bul2.2015.1720420109

Chase, R. J., and G. LaPorte. 2018. "The Next Generation of Crime Tools and Challenges: 3D Printing." *National Institute of Justice*, no. 279. https://www.ncjrs.gov/pdffiles1/nij/250697.pdf.

Cullinane, S. 2018. "Monkey Does Not Own Selfie Copyright, Appeals Court Rules." *CNN*. https://www.cnn.com/2018/04/24/us/monkey-selfie-peta-appeal/index.html.

Hall, N. 2016. "IP Losses to Top $100 Billion in 2018." *3D Printing Industry*. May 30, 2016. https://3dprintingindustry.com/news/ip-losses-top-100-billion-2018-80821.

Hornick, J. 2015. "3D Printing and IP Rights: The Elephant in the Room." *Santa Clara Law Review* 55, no. 4: 801–18. https://digitalcommons.law.scu.edu/lawreview/vol55/iss4/1.

Hornick, J., and R. Wells. 2016. "EnvisionTEC Patent Suit against Formlabs: An IP Perspective." 3Dprint.com. October 3, 2016. https://3dprint.com/151221/envisiontec-formlabs-ip-view.

Malaty, E., and G. Rostama. 2017. "3D Printing and IP Law." *WIPO Magazine*. http://www.wipo.int/wipo_magazine/en/2017/01/article_0006.html.

Morelle, R. 2013. "Working Gun Made with 3D Printer." BBC News. May 6, 2013. http://www.bbc.com/news/science-environment-22421185.

Naftulin, J. S., E. Y. Kimchi, and S. S. Cash. 2015. "Streamlined, Inexpensive 3D Printing of the Brain and Skull." *PLoS ONE* 10, no. 8: e0136198. DOI: 10.1371/journal.pone.0136198.

US Food and Drug Administration. 2017a. "FDA'S Role in 3D Printing." https://www.fda.gov/MedicalDevices/ProductsandMedicalProcedures/3DPrintingofMedicalDevices/ucm500548.htm.

———. 2017b. "Statement by FDA Commissioner Scott Gottlieb, M.D., on FDA Ushering in New Era of 3D Printing of Medical Products; Provides Guidance to Manufacturers of Medical Devices." https://www.fda.gov/NewsEvents/Newsroom/PressAnnouncements/ucm587547.htm.

———. 2017c. "Technical Considerations for Additive Manufactured Medical Devices Guidance for Industry and Food and Drug Administration Staff." https://www.fda.gov/downloads/MedicalDevices/DeviceRegulationandGuidance/GuidanceDocuments/UCM499809.pdf.

———. 2018. "Classify Your Medical Device." https://www.fda.gov/MedicalDevices/DeviceRegulationandGuidance/Overview/ClassifyYourDevice/ucm2005371.htm.

US Department of Health and Human Services. 2013. "Summary of the HIPAA Privacy Rule." https://www.hhs.gov/hipaa/for-professionals/privacy/laws-regulations/index.html.

World Intellectual Property Organization. 2011. "What Is Intellectual Property?" http://www.wipo.int/edocs/pubdocs/en/intproperty/450/wipo_pub_450.pdf.

Chapter Three

A Literature Review on 3D Printing

3D printing services have seen a rise in popularity in libraries of all types. Public and academic libraries took an early leap, began implementing the service, and aided other libraries by sharing their experiences. The Fayetteville Free Library became the first public library to offer 3D printing in 2011, and in 2012, the DeLaMare Science and Engineering Library of the University of Nevada, Reno, began offering 3D printing in its library (Zuberbier, Agarwala, Sanders, and Chin, 2016).

Amanda Goodman, director at large for a library system, has worked over the years to compile a running list of libraries with 3D printers. With the data she collected, Goodman created a Google Map with pins highlighting libraries across the world with 3D printers. In her most recent data upload, the total number of libraries with 3D printing services has reached 864 (Goodman, n.d.). With so many libraries now 3D-printing, there is an abundance of information on how best to start this service.

IMPLEMENTING A 3D PRINTING SERVICE

Early research focused on the needs for 3D printing in the library, as well as the method in which to set up and offer this service. Beyond partners and collaborators with the service, it is useful to know some basics to have on hand prior to starting a service so you know what to order and can add to the budget. Rachael Elrod (2016) describes the various elements to consider when starting a 3D printing service, including plastic bags for delivering models, tools for postprocessing and maintenance, and such other unique elements as a 3D-printer cart for a mobile workstation.

Northern Arizona University's Cline Library opened a MakerBot Innovation Center in 2015, with twenty 3D printers and scanning equipment. The

Cline Library's center is unique in that it was set up in a matter of months. The library staff made recommendations for libraries to avoid and other items to have established before opening a service for business. One recommendation is to develop a workflow and train multiple staff members to ensure that several people are available to assist during busy times. Another recommendation is to take a field trip to another 3D print lab. Field trips to other print labs were also useful for the Ruth Lilly Medical Library prior to starting their 3D printing service, as they provided valuable insights into the current demand on campus and any challenges that may be encountered.

In a review of their public library's 3D printing service, Jason Griffey (2017) discusses the benefits of using 3D-printer-management software. Libraries can vary by type and have different demands and uses, but the benefits of using such software are easily translatable to all libraries with multiple printers. Such benefits include starting prints over a wireless network, receiving notifications on print status, and checking on print status via webcam.

When visiting early library 3D print labs, Barbara Jones, director of the American Library Association's Office for Intellectual Freedom, described the "seat-of-the-pants implementation" as both "exciting and scary" (2015). Jones highlights the apprehensions of librarians when it came to 3D printing, such as the difficulties determining what could and could not be printed. This decision was not based on whether the printers were capable but whether policies on a model's "content" were too strict or too relaxed. By creating policies early on, these worries can be lessened. To help develop policies, it is important to really understand the library's role in 3D printing. Jones says it well when describing 3D printers in libraries: "The goal of 3D printing in libraries will not be to create the next artificial heart. . . . A library 3D printing service could easily inspire future medical technicians or entrepreneurs and give them a head start on creating that heart in a future career in a highly specialized medical laboratory" (Jones, 2015). As long as libraries understand the purpose of their 3D printers, they will be better able to develop policies to support their goals and defend their decisions and the service itself. Different libraries have different goals for their printers, and the strictness of policies should reflect these goals.

Beyond policies, equipment, and training, another aspect to consider when starting a 3D printing service is elements outside the library. Developing partnerships and relationships both within and outside your library are important to keep a 3D printing service running, as librarians at Northern Arizona University's Cline library noted, "Partnerships were the key to our success" (Crum, 2017). Faculty and staff are an obvious choice to partner with when developing a 3D printing service, but it is also important to think outside the box and consider nonacademic partners. The W. E. B. Du Bois Library at the University of Massachusetts, Amherst, opened a MakerBot Innovation Center with fifty 3D printers available for students, staff, faculty, and the community (Enis, 2015). The library's former director, Jay Schafer, describes benefits of collaborating with businesses near the university to foster innovation and entrepreneurship, as well as educating students in creating business plans and forming public–private partnerships (Connare, 2015).

3D PRINTING IN PUBLIC AND SCHOOL LIBRARIES

Though public and school libraries have different communities they are helping, it is important to see what all libraries are doing with 3D printing and what brings challenges and successes to the service. Public libraries were the first to embrace 3D-printing technology and, as such, have more history with the service and knowledge of potential pitfalls. Public libraries typically include printers and scanning equipment, as well as dedicated computers with software for designing models, and may or may not require librarian supervision. In Dayton, Ohio, five public libraries received 3D printers. They required patrons to limit print times to under six hours and provide models on a flash drive for librarians to start and monitor (McClory, 2015). Bohyun Kim (2016) also lists important details that public libraries need to review to establish a successful service to meet the needs of their community.

K–12 school libraries have also taken part in 3D printing, and with the 2015 *NMC Horizon Report K–12* supporting the expanded use of 3D printers in education, the library is an ideal place to house the machines. Additionally, MakerBot is supporting training and education, as well as supplying the machines themselves, with its MakerBot Academy initiative. The initiative plans to provide a 3D printer for every school across the country. All levels of K–12 have opportunities with 3D printing, as education librarian Elrod discovered when researching implementation of a 3D printing service in her university library. In her literature review, Elrod learned that kindergartners helped to design playground equipment and middle-schoolers printed art projects (Elrod, 2016). Additionally, librarians Cano (2015) and Moorefield-Lang (2014a; 2014b; 2015) highlight how elementary, middle, and high school libraries can use 3D printers in a variety of curriculums, including math, science, language arts, and social studies.

3D PRINTING IN ACADEMIC LIBRARIES

Libraries—public, university, and medical—share resources across a community. In academia, libraries are the antisilo and bring together groups that may not have typically interacted. Bruce Massis (2013) recognizes that libraries exist in think tanks, where students collaborate on projects and intellectual curiosity is thriving among not only students but faculty, as well. This puts libraries in the perfect position to engage their communities with 3D printing services. The DeLaMare Science and Engineering Library of the University of Nevada, Reno, recognized this multidisciplinary feature of 3D printing when they investigated and implemented the service in 2012. The library was an early pioneer in 3D printing and the first academic library to offer this service to university faculty, students, and staff, as well as the public. The first request the library received proved to be very insightful about how the service could help students: Engineering students who previously outsourced their 3D printing benefited from the timely, on-site service available in the library, where they could then continue to work on updating and revising the design in another library space. Nearly 50 percent of students used the service to support mechanical engineering coursework; art and biochemistry came in second, fulfilling 11 percent of the total use (Colegrove, 2014).

When setting up a 3D printing service in the library, it is important to consider the space options to match the intended outcomes of the service. Gillian Nowlan (2015) discusses the location of a 3D printing service area within the library and highlights the importance of a central location with an open space. Depending on whether students can use the printers, safety training is another element of the service to plan. Vincent F. Scalfani and Josh Sahib (2013) discuss setting up safety standards through training offered at workshops.

While the workshops at Scalfani's library were popular, software and printer issues tend to be troublesome and time-consuming despite the training. Authors note that initial prints from their group of users who passed training had model-design failures or problems during printing. This highlights another chief concern with starting a printing service—learning how to use the equipment. The skills needed to operate and run a 3D print lab are reviewed in Kyungwon Koh and June Abbas (2015) and Beth Williams and Michelle Folkman (2017). Skills that library staff need to successfully operate a 3D printing service include being able to learn quickly, adapt to change, collaborate, justify the service, and assist a wide range of users.

Another use of a 3D printing service in the library goes beyond students and faculty and involves the library itself. Michael Groenendyk and Riel Gallant (2013) identify an archival use of 3D-printing and scanning technology: Libraries with special collections can put their displays online and make previously fragile and untouchable objects manipulatable.

3D PRINTING IN HEALTH SCIENCES AND MEDICAL LIBRARIES

Anatomist Michelle Smith and James Jones published an article in the *Anatomical Sciences Journal* describing the benefits of 3D printing for both students and educators. They state that libraries are a perfect location for offering this service (2018).

With growing faculty interest and existing student interest, health sciences and medical libraries have carefully begun to delve into the world of 3D printing. Matthew Hoy (2013) recognizes the need for medical librarians to pay special attention to 3D printing and explains that libraries could provide a learning ground for 3D printing for medical students and professionals to take into their careers. J. Dale Prince similarly describes the benefits and uses of 3D printing in medicine and education and echoes Hoy's sentiment about libraries' involvement in using the technology to expand their service offerings.

In 2014, the National Institutes of Health began a six-month pilot program, offering a free 3D printing service to scientists. During this time, the librarians tracked the number of prints, where requestors came from (institute-wise), and printer issues. Printer maintenance and the number of replacement parts were recorded to give a good estimate of expenses. The success of the pilot program led to continuing the service, and the following year saw an increase in the number of prints and a steady stream of attendees for their 3D printing orientation. Verma Walker (2017) notes the drop in attendance for orientation after the first year. Initially, this might seem like a drop in interest of the service, but as Walker highlights, this is in part due to users already attending the orientation in 2014.

Health sciences and medical libraries also have the potential to become integrated in curriculums and coursework through their 3D printing service. Laurel Scheinfeld, health sciences librarian with Touro College, has successfully integrated in an occupational therapy pilot program, where the library plays a key role in providing the 3D printing service to allow students to design and create assistive devices. Librarians work with faculty and students and are embedded in the course to teach students about the printers and the processes involved with 3D models. In addition to their role in supporting this course, librarians play a role in fostering interdisciplinary education, as engineering students work with occupational therapy students to create the devices (Scheinfeld, 2018). Bridle and Schultz (2015) also note the interdisciplinary activities 3D printing generates, noting that engineering, zoology, chemistry, mathematics, and experimental psychology were just some of the departments that interacted with 3D printing within their library.

One aspect of 3D printing that is still being investigated is the potential health effects that could occur due to particulates or fumes that come from the printers. Sara Russell Gonzalez and Denise Beaubien Bennett (2014) review concerns in 3D printing that go beyond the service itself and focus on environmental safety, while still touching on staff training, marketing, and maintenance. Migraine- and asthma-prone staff reported no rise in symptoms with the start of the service; however, other studies have found different results.

HEALTH EFFECTS OF 3D PRINTING

One issue with 3D printing involves particulates that may be emitted. While some printers and materials note the need for ventilation, it is those printers that do not stipulate this need that can pose problems with the service. Fumes and particulates and their effects on staff running the printers have been studied. In one study, staff from seventeen different 3D printing companies reported experiencing respiratory problems at some point during their work week (Chan et al., 2018). The companies surveyed included businesses, universities, and libraries. It was found that longer exposure to the 3D printers increased the risk of a respiratory illness. In one case study, a staff member using multiple FDM printers began experiencing asthma symptoms within two weeks of use in a confined space (House, Rajaram, and Tarlo, 2017). The material used was ABS filament, and after changing to PLA and adjusting how many printers operated at the same time, the staff member's symptoms improved. One researcher highlights the fact that respiratory problems can occur even with traditional office equipment and notes that respiratory symptoms also are prevalent for some office workers using photocopies (Karimi et al., 2016).

One study compares exposure to ABS and PLA plastics for only one hour (Gumperlein et al., 2018). In this study, ABS was found to have an odor and increased levels of exhaled nitric oxide, believed to be caused by "eosinophilic inflammation" resulting from inhaling ultrafine particulates emitted during printing. When comparing a single printer in a larger ventilated space to continued-use printers in a smaller unventilated space, larger amounts of ultrafine aerosols were emitted and lingered after more than twenty hours (Steinle, 2016).

Filament type has also been investigated more thoroughly. ABS filament has repeatedly been found to have worse emissions of particulates (Wojtyla, Klama, and Baran, 2017), with color also found to be a factor in rate of emissions. Natural filament lacking color alteration was found to have the highest emissions, with black having the least (Yi et al., 2016). Wood-infused PLA was also discovered to have a lower amount of emission (Vance, Pegues, Van Montfrans, Leng, and Marr, 2017). Nylon, brick, and ABS all had the highest amounts of volatile organic compounds (Azimi, Zhao, Pouzet, Crain, and Stephens, 2016).

MAKERSPACES

Another consideration of having a 3D printing service in libraries is developing it as part of a makerspace. 3D printing and makerspaces are close to synonymous due to the creative and inventive endeavors taken by the users. Makerspaces act as a hub for making, or creating, with 3D printing a spoke in the wheel, working in conjunction with other services to support users' work. Setting up a 3D printing service in itself is a large task and requires a lot of time to set up, market, and then run. Setting up a makerspace takes even more time, as more elements need to be considered and potentially more service offerings made available. Additionally, as the name suggests, makerspaces need more space than a 3D printing lab might require. This space might also need special considerations for equipment, such as ventilation.

Susana Nascimento and Alexandre Polvora (2018) researched maker culture and describe how the low cost of 3D printers and availability of open-sourced software have fueled the maker movement. Furthermore, they believe that the maker culture has the "potential to turn users into active designers, producers, creators and distributors of knowledge, tools and machines" (943). At the University of Tennessee, faculty integrated their makerspace into nursing rotations where students spent time at the hospital and then at the makerspace. They worked with engineering students to design devices to assist in potential patient care. On review of the program, faculty found that students created models that could benefit patient care in the future.

Whether libraries choose to launch a 3D printing service or go bigger and open a makerspace to house their 3D printers, the same issues and problems can occur. Reviewing makerspaces and maker culture can provide insight into potential service opportunities and give a good idea of the kinds of users who will be interested in the service. More resources on the maker movement appear in chapter 12.

REFERENCES

Azimi, P., D. Zhao, C. Pouzet, N. E. Crain, and B. Stephens. 2016. "Emissions of Ultrafine Particles and Volatile Organic Compounds from Commercially Available Desktop Three-Dimensional Printers with Multiple Filaments." *Environmental Science and Technology* 50, no. 3: 1260–68. DOI: 10.1021/acs.est.5b04983.

Bridle, O., and K. Schultz. 2015. "3D Printing and Scanning: New Ways to Engage with Students and Researchers." *Proceedings of the IATUL [International Association of Scientific and Technological University Libraries]Conferences*. Paper 2. http://docs.lib.purdue.edu/iatul/2015/future/2.

Cano, L. M. 2015. *3D Printing: A Powerful New Curriculum Tool for Your School Library*. Santa Barbara, CA: Libraries Unlimited.

Chan, F. L., R. House, I. Kudla, J. C. Lipszyc, N. Rajaram, and S. M. Tarlo. 2018. "Health Survey of Employees Regularly Using 3D Printers." *Occupational Medicine* 68, no. 3: 211–14. DOI: 10.1093/occmed/kqy042.

Colegrove, P. 2014. "Making It Real: 3D Printing as a Library Service." *EDUCAUSE Review*. October 27, 2014. https://er.educause.edu/articles/2014/10/making-it-real-3d-printing-as-a-library-service.

Connare, C. 2015. "MakerBot Innovation Center Grand Opening March 26." *UMass Amherst Libraries*. https://www.library.umass.edu/news/events/makerbotopening/.

Crum, J. A., J. Hillock, B. J. Johnson, and K. L. Schmand. 2017. "NAU's Cline Library: How We Launched Our MakerBot Innovation Center in No Time." *Computers in Libraries* 37, no. 2.

Elrod, R. 2016. "Tinkering with Teachers: The Case for 3D Printing in the Education Library." *Education Libraries* 39, no. 1.

Enis, M. 2015. "UMass Amherst Library Opens 3-D Printing Innovation Center." *Library Journal*. https://lj.libraryjournal.com/2015/03/technology/umass-amherst-library-opens-3d-printing-innovation-center.

Gonzalez, S. R., and D. B. Bennett. 2014. "Planning and Implementing a 3D Printing Service in an Academic Library." *Issues in Science and Technology Librarianship*. DOI:10.5062/F4M043CC

Goodman, A. L. n.d. "Map of 3D Printers in Libraries." Accessed June 3, 2018. http://www.amandagoodman.com/3d.

Griffey, J. 2017. "3-D Printers for Libraries, 2017 Edition." *Library Technology Reports* 53, no. 5: 1.

Groenendyk, M., and R. Gallant. 2013. "3D Printing and Scanning at the Dalhousie University Libraries: A Pilot Project." *Library Hi Tech* 31, no. 1: 34–41. https://doi.org/10.1108/07378831311303912.

Gumperlein, I., E. Fischer, G. Dietrich-Gumperlein, S. Karrasch, D. Nowak, R. A. Jorres, and R. Schierl. 2018. "Acute Health Effects of Desktop 3D Printing (Fused Deposition Modeling) Using Acrylonitrile Butadiene Styrene and Polylactic Acid Materials: An Experimental Exposure Study in Human Volunteers." *Indoor Air* 28, no. 4: 611–23. DOI: 10.1111/ina.12458.

House, R., N. Rajaram, and S. M. Tarlo. 2017. "Case Report of Asthma Associated with 3D Printing." *Occupational Medicine* 67, no. 8: 652–54. DOI: 10.1093/occmed/kqx129.

Hoy, M. B. 2013. "3D Printing: Making Things at the Library." *Medical Reference Services Quarterly* 32, no. 1: 93–99. DOI: 10.1080/02763869.2013.749139.

Jones, B. 2015. "3D Printing in Libraries." ASIS&T: Association for Information Science and Technology. https://www.asist.org/publications/bulletin/oct-15/3d-printing-in-libraries.

Karimi, A., S. Eslamizad, M. Mostafaee, Z. Momeni, F. Ziafati, and S. Mohammadi. 2016. "Restrictive Pattern of Pulmonary Symptoms among Photocopy and Printing Workers: A Retrospective Cohort Study." *Journal of Research in Health Sciences* 16, no. 2: 81–84.

Kim, B. 2016. "Building Blocks of an Innovation Space: Field Reports." *Library Journal*. February 16, 2016. https://www.libraryjournal.com/?detailStory=building-blocks-of-an-innovation-space-field-reports.

Koh, K., and J. Abbas. 2015. "Competencies for Information Professionals in Learning Labs and Makerspaces." *Journal of Education for Library and Information Science* 56, no. 2: 114–29.

MarketWatch. 2015. "As Libraries across the U.S. Embrace Desktop 3D Printing, UMass Amherst Opens First Large-Scale MakerBot Innovation Center at a University Library." Press release, March 24, 2015. https://www.marketwatch.com/press-release/as-libraries-across-the-us-embrace-desktop-3d-printing-umass-amherst-opens-first-large-scale-makerbot-innovation-center-at-a-university-library-2015-03-24.

Massis, B. E. 2013. "3D Printing and the Library." *New Library World* 114, nos. 7–8: 351–54. https://doi.org/10.1108/NLW-03-2013-0030.

McClory, E. 2015. "3D Printing Available at 5 Library Branches." *Dayton* [OH] *Daily News*, July 13, 2015.

Moorefield-Lang, H. M. 2014a. "3-D Printing in Your Libraries and Classrooms." *Knowledge Quest* 43, no. 1: 70–72.

———. 2014b. "Makers in the Library: Case Studies of 3D Printers and Maker Spaces in Library Settings." *Library Hi Tech* 32, no. 4: 583–93. https://doi.org/10.1108/LHT-06-2014-0056.

———. 2015. "Change in the Making: Makerspaces and the Ever-Changing Landscape of Libraries." *TechTrends* 59, no. 3: 107–12. https://doi.org/10.1007/s11528-015-0860-z.

Nascimento, S., and A. Polvora. 2018. "Maker Cultures and the Prospects for Technological Action." *Science and Engineering Ethics* 24, no. 3: 927–46. DOI: 10.1007/s11948-016-9796-8.

Nowlan, G. A. 2015. "Developing and Implementing 3D Printing Services in an Academic Library." *Library Hi Tech* 33, no. 4: 472–79. https://doi.org/10.1108/LHT-05-2015-0049.

Prince, J. D. 2014. "3D Printing: An Industrial Revolution." *Journal of Electronic Resources in Medical Libraries* 11, no. 1: 39-45. https://doi.org/10.1080/15424065.2014.877247

Scalfani, V. F., and J. Sahib, J. 2013. "A Model for Managing 3D Printing Services in Academic Libraries." *Issues in Science and Technology Librarianship*. DOI:10.5062/F4XS5SB9

Scheinfeld, L. 2018. "3D Printing Presentation at ACRL New England Chapter Annual Conference." Stony Brook University Libraries. Association for College and Research Libraries. May 4, 2018. https://library.stonybrook.edu/3d-printing-presentation-at-acrl-new-england-chapter-annual-conference.

Smith, M. L., and J. F. X. Jones. 2018. "Dual-Extrusion 3D Printing of Anatomical Models for Education." *Anatomical Sciences Education* 11, no. 1: 65–72. DOI: 10.1002/ase.1730.

Steinle, P. 2016. "Characterization of Emissions from a Desktop 3D Printer and Indoor Air Measurements in Office Settings." *Journal of Occupational and Environmental Hygiene* 13, no. 2: 121–32. DOI: 10.1080/15459624.2015.1091957.

Vance, M. E., V. Pegues, S. Van Montfrans, W. Leng, and L. C. Marr. 2017. "Aerosol Emissions from Fuse-Deposition Modeling 3D Printers in a Chamber and in Real Indoor Environments." *Environmental Science and Technology* 51, no. 17: 9516–23. DOI: 10.1021/acs.est.7b01546.

Walker, V. 2017. "Implementing a 3D Printing Service in a Biomedical Library." *Journal of the Medical Library Association* 105, no. 1: 55–60. DOI: 10.5195/jmla.2017.107.

Williams, B. F., and M. Folkman. 2017. "Librarians as Makers." *Journal of Library Administration* 57, no. 1: 17–35. DOI: 10.1080/01930826.2016.1215676.

Wojtyla, S., P. Klama, and T. Baran. 2017. "Is 3D Printing Safe? Analysis of the Thermal Treatment of Thermoplastics: ABS, PLA, PET, and Nylon." *Journal of Occupational and Environmental Hygiene* 14, no. 6: D80–85. DOI: 10.1080/15459624.2017.1285489.

Yi, J., R. F. LeBouf, M. G. Duling, T. Nurkiewicz, B. T. Chen, D. Schwegler-Berry, M. A. Virji, and A. B. Stefaniak. 2016. "Emission of Particulate Matter from a Desktop Three-Dimensional (3D) Printer." *Journal of Toxicology and Environmental Health, Part A* 79, no. 11: 453–65. DOI: 10.1080/15287394.2016.1166467.

Zuberbier, D., R. Agarwala, M. Sanders, and R. Chin. 2016. "An Academic Library's Role in Improving Accessibility to 3-D Printing." American Society for Engineering Education, 2016 ASEE Annual Conference and Exposition. https://www.asee.org/public/conferences/64/papers/15375/view.

Chapter Four

3D Printing Service Survey

On April 10, 2018, a survey was sent to the Medical Library Association LISTSERV, requesting responses from members about their 3D printing service or thoughts on a 3D printing service. The survey had nine questions that asked respondents to provide some background information on their library and 3D printing service status, followed by questions on the biggest challenges and struggles they've encountered with either starting the service or with their existing service. Questions about the details of their service were also included, such as what they charge for 3D printing and what type of postprocessing support they offer. The final questions asked about requests they receive and if they were integrated in any curriculums or other coursework. An open-ended question finished the service, asking respondents if they had any other information to share about their service. The survey remained open until the end of May and received seventeen responses. The following is a summary of those responses and what was learned about 3D printing services from members of the Medical Library Association.

A majority of responses came from academic health sciences libraries, making up almost 65 percent of the total responses. Hospital libraries came in second, with almost 25 percent of the responses. Academic medical libraries filled the remaining 10 percent. Respondents overwhelmingly described funding as the biggest challenge of implementing the service, with the cost of printers, materials, and supplies and ongoing funding listed as concerns. Additional issues included training and perceived lack of technology skills. Limited staff and time also were mentioned, as well as difficulty getting support from other departments and buy-in from faculty. Implementing policies was also noted as a challenge. The following are some of the challenges respondents cited as a hurdle in establishing their 3D printing services:

> "Cost and skills training—a good printer is expensive, and building the skills to properly take advantage of it is very time-consuming."

> "Staffing, equipment selection and troubleshooting, material choices, IT support sources."

> "Learning about 3D-printing technology, designing an appropriate space, and providing professional development to a librarian and student workers."

> "Support from other departments on campus. Support from administration/IT."

> "Interest. We knew we could get into 3D printing, but wanted to make sure that if we bought the stuff, people would use it."

When asked about struggles with existing 3D printing services, maintenance was found to be the biggest issue:

> "Our big printer has regular problems because we use it very heavily. When it stops and knocks over models, we have to use more material and restart projects, which can cause big unexpected delays and costs."

> "Printers break. Requests are overwhelming. Need more staff and space to continue growing."

Other issues libraries encountered included keeping up with requests, limited time, staff availability, cost to continue the service, promotion, awareness, and justifying the service.

Many libraries operating a 3D printing service did not charge for the service. Of the libraries with active 3D printing services, 40 percent charged an average of ten cents per gram, with the other 60 percent offering the service at no cost, as piloting the service for free is a common practice to draw interest. A review of postprocessing included with the service yielded mixed results. Postprocessing models involves removing support material, which is supposed to break away, but those who have experience with 3D printing know that the supports do not always break cleanly away and may require additional postprocessing such as sanding and smoothing models. In some cases, some postprocessing may involve painting models. Respondents in the survey were split almost 50/50 in regard to offering this additional service. Funding, staffing, and a sometimes-time-consuming service can be taxing and take up more time than prepping the printers and starting the prints themselves.

Respondents were asked about the requests they've printed. Their answers revolved around anatomical models with unique elements, which illustrates how students' learning needs are different and how faculty's instructional needs vary:

Figure 4.1. Main challenges in 3D printing services.

> "A faculty member requested to have a brain printed that he modified to look like one that had undergone stroke for students and patients to see."

> "Faculty members are interested in hearts and difficult airways. Students are interested in kidneys, pediatric skeleton, interior views of the skull, liver, adult skeleton, and bones of the hand and foot that are larger than scale."

> "Heart models in four slices, half-skulls, reproductions of a particular sculpture that is beloved by our campus, vertebrae (LOTS of them), hyoid bones and the connected neck, teeth, dental tools, assistive equipment, research pieces, lizard heads, etc. They've been requested by all three—we encourage it. The most unusual things are mostly faculty, but not all."

> "Anatomical structures, machine parts, invention prototypes, etc., requested by students, faculty, staff, and clinicians."

To conclude the survey, libraries were asked about their involvement in curriculums and any coursework. Roughly 65 percent of libraries responded that they were integrated in some way with student work. Comments included:

> "Yes. Some members of the library staff worked with teams of students to apply for awards to do certain research projects based on 3D printing. We also provide models for specific courses to use."

> "1. Teaching in a first-year symposium week; 2. class project for OT students (students need to locate and print an OT device); 3. graded, two-week fourth-year elective."

The next chapter features some of the libraries who responded in the survey and volunteered to write a case study. Case studies about their libraries' 3D printing offer insights into their services. As training and developing the skills necessary to run a 3D printing service were cited as obstacles and challenges, later chapters address some basics of 3D printing and how to find models, edit models, and troubleshoot common problems. Additionally, recommended resources are available for continued support in 3D printing. To review all survey responses, please see Appendix A.

Chapter Five

Case Studies

THE A. T. STILL MEMORIAL LIBRARY 3D PRINT SHOP

Susan E. Swogger, MLIS Librarian
Debra Loguda-Summers, 3D Print Shop and Public Services Manager
A. T. Still Memorial Library, A. T. Still University, Missouri

The Setting and Goals

The A. T. Still Memorial (ATSM) Library of A. T. Still University (ATSU) has a bifurcated location, with separate but jointly administered campuses in both Missouri and Arizona. ATSU is a small, nonprofit, private, graduate university made up of two osteopathic medical schools, two dental schools, and a variety of graduate health sciences programs, both on campus and online. Each library location offers 3D print shop services to its local campus but with a larger footprint and investment at the A. T. Still Memorial Library in Missouri (ATSM Library–Missouri). This case study focuses primarily on the 3D print shop service as it has developed at the library branch in Missouri.

The impetus for establishing 3D printing as a generally available service came from both a strong administrative champion—the university president—and from library interest in expanding its services portfolio. The university wanted to use 3D printing to engage student and faculty interest in a rapidly expanding technology, important to the health sciences, to produce 3D models customized to university instructional and research aims and to provide opportunities for student health scientists to explore new ways to use models in education and clinical practice. The two library branches were a natural fit for managing and housing the new 3D printing initiative: We had staff interest, an already-established interdisciplinary space for all programs, and physical and managerial capacity.

Initial Equipment and Outreach

In early 2015, the university purchased a Stratasys MOJO 3D printer and a WaveWasher model cleaner and dedicated a computer for each campus library at a startup cost of approximately six thousand dollars per location. This printer was chosen for its detailed printing capability and the easy cleanability of the models produced. The library heavily promoted the newly available 3D printing service through advertisement on its main website; the creation of a detailed LibGuide to share current 3D models and processes available for printing; communication and direct contact with departmental liaisons; and an enthusiastically received contest in 2017. It continues to promote it by offering regular news stories about interesting recent 3D prints and available training or lectures in its newsletter and on the university blog and hopes to hold contests annually. We were able to offer a generous prize to winners on each campus for our initial contest, and our user community responded with a wide variety of projects focused on anatomy, instruction, assistive technology, and more, exactly as we had hoped. The contest was a particularly effective outreach tool because we were able to involve faculty and staff from several areas in promoting it directly, as well as in providing funds to support it.

University support allowed us to offer the service at no charge to the entire university community for anything with an educational, research, or institutional support purpose. As students generally lack excess funds, we expected this to be a key factor in coaxing them to try out the new service, a viewpoint later confirmed in our first systematic evaluation effort.

Initial interest in and curiosity about the new service was strongest on the Missouri campus. In its first year, the ATSM Library–Missouri partnered with students to apply for a five-thousand-dollar in-house grant to purchase the equipment needed for the 3D scanning capability in its 3D print shop. The application was successful, and we purchased a NextEngine 3D laser scanner, an iMac computer, and Osirix MD software. The incorporation of this additional equipment and software allows us to create 3D models with the laser scanner or use the Osirix MD software to create files from CT scans, PET scans, and MRI scans when free files are unavailable. By actively partnering with students on the grant that funded its purchase, we were able to closely engage not only the students directly involved but also their peers and supervising faculty. Allowing them full agency in the grant-writing process, implementation of their intended project, and use of the equipment visibly increased their investment in participating in and sharing news about the 3D printing initiative.

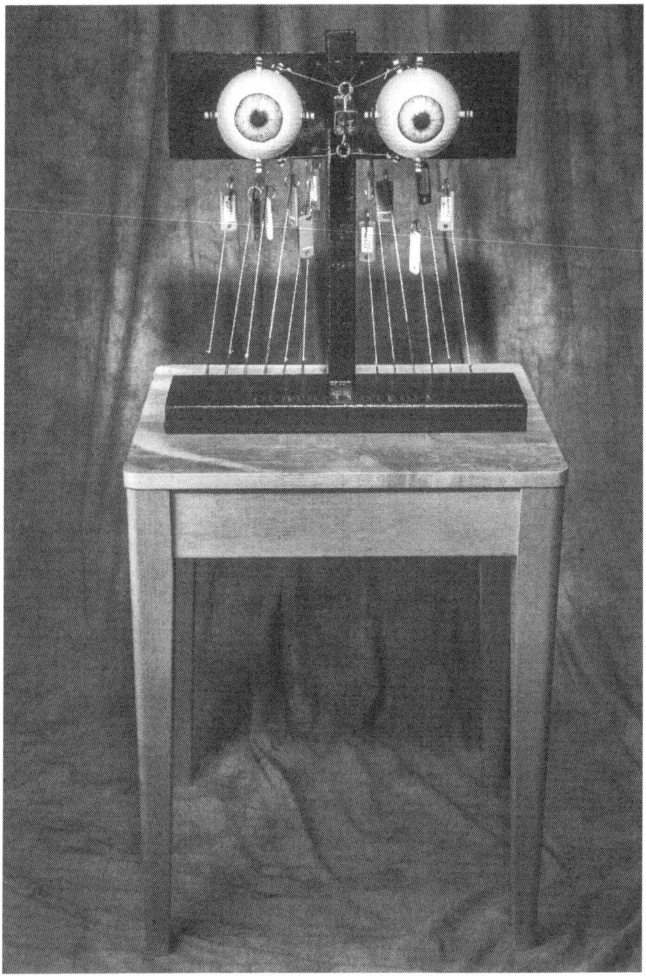

Figure 5.1. ATSM Library–Missouri 3D printing contest winner: the ophthalmotrope. Designers: Jeremy J. Houser, PhD; Baydon Hilton, OMS II; and Blake Stringham, OMS II, of A. T. Still University. *Kelly Rogers, A. T. Still University.*

First Evaluation

In 2017, ATSM Library–Missouri entered into our first formal evaluation of the program, sending a survey to all ATSU students and faculty about their views on our 3D printing services, regardless of location. The survey asked six questions about their opinions of the benefits of 3D models to anatomical learning; their willingness to pay for this service; their interest in printing larger models; and if or how they might use 3D printing in clinical practice.

We received 318 responses, with 288 including comments. Our initial conclusion was that the ATSU community does value the 3D service and does perceive educational and clinical benefits to its availability. However, they did not express a willingness to pay any more than a very nominal fee for individual models; many explicitly stated that their general tuition should include it as it was seen as a learning tool. It was a clear implication that our ability to subsidize it was strongly connected to its success and wide adoption.

Expansion

When demand continued to grow past the Mojo's production and size capacity, the ATSM Library–Missouri was able to use that fact and the results of the survey (interest in larger models was nearly universal) to make a case for expanding its capacity again. The library used its endowment fund as well as administrative funding to purchase a larger-capacity printer and cleaner—a Stratasys F170 printer capable of creating models up to ten inches and a large SCA 1200 HT support cleaner. Stratasys makes several educational-package offers available, including this one, available to us for $32,000. While the Mojo can sit on a desk and share an outlet with a computer, the F170 has much greater infrastructure needs—it weighs approximately five hundred pounds, is sixty-four inches tall by thirty-four inches wide, and requires a dedicated wall outlet with a specialized heavy-duty surge protector. Acquiring and properly housing one requires a substantial investment beyond the start-up costs.

Challenges

As with any new or ongoing service, we have had our difficulties; sufficient funding for supplies and maintaining designated staff are the two main issues. Our printing costs typically range from fourteen dollars per cubic inch to twenty-one dollars per cubic inch, depending on printer and materials. These costs include printing material, support material, printing bases, and cleaning materials. At this time, the university budgets ten thousand dollars per year for supplies; based on past levels of use, we expect the budget's needs to triple to thirty thousand dollars a year for supplies as requests for projects increase. At present, this does not come directly out of library funding but is instead directly funded by university administration and managed by library staff. While this has been generous, costs are increasing, and funding is not guaranteed past another year. When the university administration's commitment ends, funding materials at present levels will be challenging.

This dedicated funding does *not* include the cost of library staff time; as 3D printing is time-consuming, this is a significant impact on the library. The library was initially able to rely on various interested students and staff from other departments for part of the staffing and expertise needed, but this quickly (and predictably) ceased to be sustainable. The existing position of ATSM Library–Missouri public services manager was rewritten to include management of the 3D print shop, but no funds were available to add additional permanent staff full time or to reassign work sacrificed to add to the print shop capacity. It is not difficult to find student workers interested in working with the print shop, but teaching them requires serious investment of staff time, and it is crucial to maintain the expertise of the permanent library staff for all equipment for the sustainability of the program. Fortunately, we were able to rearrange positions on each campus to include 3D printing, but it did require readjustment of priorities for some other existing tasks.

Infrastructure needs are also a challenge, especially with the addition of the larger-capacity printer. As mentioned previously, the size and special requirements of the Statasys F170 need much more than a simple desktop 3D printer, such as our Mojo. In addition to its initial cost, we funded the installation of a dedicated electrical and internet line and the purchase of a high-voltage power-surge protector and requisitioned a dedicated IP address isolated from the rest of the campus network. The Statasys F170 also requires the regular purchase of special printer heads for each of the materials used as they are not interchangeable and last for merely 1,500 hours of printing. As these cost approximately $930 per head, an unplanned jump in printing requests can wreak havoc on a budget with little leeway. Our heavy printing volume and the occasional unexpectedly poor-quality file have also resulted in the need for time-consuming recalibration, responding to error messages, restarting the machine, reprinting complicated models, and more. One of the reasons these issues have been so frequent has been the lack of available vendor training for the very new machine; a more established machine might prove less problematic (though we love our F170 and would not give it up).

Current Production and Processes

In our view, our successes with this program have been exceptional despite the challenges. By supplying free 3D printing services to our students and faculty and embedding ourselves into several of the department's research and curriculum projects, we have more than doubled the amount of 3D printing requests each year. The ATSM Library–Missouri printed 130 models in 2016, 429 models in 2017, and 972 models in 2018 through April. The latter number is unusually high due to a special project involving many small pieces, but even without this outlier, the requests are on track to double again by the end of the year. Students and faculty alike express interest, are aware of and willing to approach the library to use this service, and generally understand its uses in health sciences education. Any event, such as a lecture, contest, or display, where the university offers focused 3D printing is always well attended.

At this point in the development of the technology, it is possible to print 3D models from metals, glass, wood, and even food, with the right printer and materials. Each print may require a different printer, materials, material density, and cleaning method. We prefer to use acrylonitrile butadiene styrene (ABS), an oil-based plastic, though we are also able to print with polylactic acid (PLA), an organic thermoplastic. The vast majority of our models are printed with ABS. The durability of this material allows the students or faculty to paint, drill, cut, and otherwise adjust to get exactly the models they need, and its ease of cleaning cuts down on the library staff's investment of time. As with all 3D-printing materials, ventilation, storage, and cleaning solution are key. Our 3D print shop has an external door that can be opened as needed to vent any fumes, as well as a Mammoth Ionic ion and ozone air purifier.

As more and more students see some of the more exciting special projects, more of them begin to place orders for items they need for study. As such, we have developed a formal request process involving a form, a brief meeting with the 3D print shop staff to discuss specific model needs, further consultation if the model request is unusual or high need, and a dedicated pick-up and retrieval space.

In order to reduce the request level only to those that are appropriate and necessary and to save ourselves from liability, we offer these limits:

1. The model must for be for educational or research use only.
2. The model cannot be prohibited by local, state, or federal law.
3. The model cannot be unsafe, harmful, or dangerous or pose an immediate threat to the well-being of others.
4. The model cannot be obscene or otherwise inappropriate.
5. The model cannot be in violation of another's intellectual property rights. For example, the printers may not be used to reproduce material that is subject to copyright, patent, or trademark protection.

Thus far, no students have requested anything that would violate any of these principles or pushed back when we required further consultation, and our very simple procedures and guidelines for these transactions have been sufficient.

Example Models and Projects

The library will print any basic project requested at no charge to students—if the project meets educational, research, or institutional aims. Departments and students purchase the materials themselves for larger-scale special projects, with the library providing its staff time and expertise at no charge to them. Having the departments purchase the supplies needed for their more unusual and resource-consuming projects decreases the cost of supplies in our budget and helps make it possible to subsidize the more normal small-scale projects that we print at no charge. The following sections offer examples of projects in each of our focal areas.

Models for Education

Many students request models of various anatomical structures for specific curricular modules. The students use the models to better understand and memorize structures and are able to keep them to share later with patients in practice. Some of the most requested models are a four-section heart model (see figure 5.2) acquired from Embodi3D (www.embodi3d.com), half- and whole skulls acquired from Thingiverse (https://www.thingiverse.com), lumbar and cervical vertebra models created by A. T. Still University's Academic Technology Department using our laser scanner, and sets of teeth or jaws acquired from various sources. Faculty and students have also printed educational toys for studying, large numbers of models for use in the classroom, supports and tools for use in the classroom and with the disabled, and more. Most applicants request one to three projects over the course of their studies, but some have requested considerably more.

Models for Research

Research projects have ranged widely and have originated from students, staff, and faculty alike. One of the most involved included creating 3D cervical, lumbar, sacrum, and pelvis models that were then encased in human-quality ballistics gel (see figure 5.3); these will be tested as user-friendly, low-cost models to assist medical students in developing the skills required to place injections into these areas. Another research project involved the reproduction of multiple scanned reptile heads in sizes ranging from 100 percent to 800 percent of the original in order to study the biophysics of sound localization in reptiles. Library involvement in these projects varies; one of the authors of this case study is heavily involved in the ballistics-gel project, and the other simply provides 3D printing services.

Figure 5.2. Four-section heart, common educational study model request. *Kelly Rogers, A. T. Still University.*

Models for Institutional Support

The library has also been able to use the 3D print shop to support institutional projects. For example, we worked with our affiliated Museum of Osteopathic Medicine to scan a 120-year-old, life-sized plaster model of the hand of Dr. Andrew Taylor Still, founder of both this university and the field of osteopathic medicine, which we then reproduced ourselves using our large F170 printer for a traveling exhibit (see figure 5.4). As the original is both fragile and beloved by the university, being able to re-create it in plastic was highly valuable to our institution.

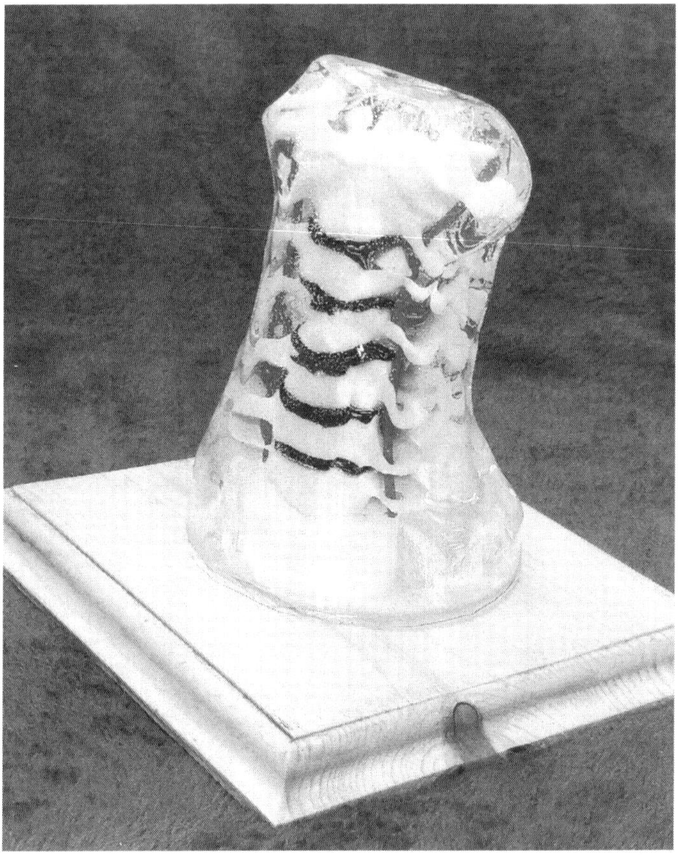

Figure 5.3. Printed cervical vertebrae model encased in ballistics gel. ***Debra Loguda-Summers.***

Next Steps

Active fiscal and outreach support from university administration to allow free experimentation by a broad segment of the campus community was key to the initial success of our library-based 3D print shop. As the funding guarantee was only for a limited period of time, we must continue to promote and extend this service if we wish to fully integrate it into university life and either attract more funding or determine the fee level students find acceptable. We intend to extend the benefits of this service to our online student population by developing a process for printing and delivering models to them. We are also actively seeking to be written into researchers' grants and department project-funding plans; funding from researchers for special pro-

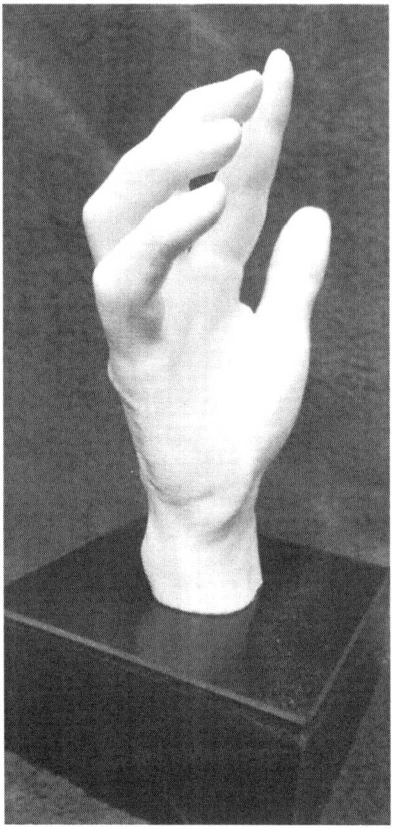

Figure 5.4. Reproduction of Dr. A. T. Still's hand. ***Debra Loguda-Summers.***

jects is allowing us to plan for continuations of some of the existing special projects and to attempt new ones.

DEVELOPING A LIBRARY MAKERSPACE

Nichelle Mack
Baptist College of Health Sciences

Baptist College of Health Sciences is located in downtown Memphis, Tennessee, and is part of the Memphis Medical District. It is a neighbor of the University of Tennessee, Health Sciences Center, and Southwest Tennessee Community College. The Baptist College Library serves current students, faculty, staff, and alumni, as well as the community. The library is located in the main building on the first floor and occupies less than six thousand square feet. A library makerspace would create another space in the area where medical and health sciences professionals and students can collaborate and define or develop solutions to improve patient care.

To develop the space, the library staff cleared an unoccupied office and rearranged some furniture. The librarian (myself), also a member of the southern chapter of the Medical Library Association, applied for a grant in September 2016, seeking five thousand dollars of the National Library of Medicine Technology Improvement Award. The award was granted, which kicked off the makerspace journey. The funding dates were January to April 2017 (https://nnlm.gov/sea/funding/funded/7358).

The items purchased with the grant funds included three 3D printers, filament, an extruder, and a 3D printing tool kit. Purchasing the items was the easy phase of the makerspace journey. The most difficult phase was executing the implementation and communication plan. There were strong learning curves for using the printers, and only two professional staff were available to train.

The safety and compliance director recommended requiring training sessions before allowing the use of the room. Sessions were scheduled four times a month and as needed. The sessions were poorly attended, and the overall use of the room was low. Faculty were not willing to devote time to explore the use of the printers or create assignments requiring their use, and students wanted more immediate use after hours or during times when the trainers were not available. The connection to student learning outcomes also presented a challenge. Students were asked to apply concepts learned in coursework but were not necessary for their practical application outside the clinical environment.

In the end, the Baptist College Library created an additional learning space for students to use for group study, which is in high demand. Library staff worked with counseling services to market the makerspace as a de-stress zone, a space where students can make anything at any time during operating hours to de-stress. The room was outfitted with exercise balls, tinker toys, and anatomy coloring sheets, and mood lighting may be installed soon. The library's Twitter account was used to communicate ideas and highlight recent prints (see figure 5.5). Library student employees began using the space to make items to keep their workspace organized, such as key chains to clearly label keys and plastic hooks to hang lanyards and jackets.

To keep the space in constant use, library staff intend to work with faculty to create a research award for students to use the 3D printers to design a solution to a problem that they have experienced during their clinical practicums. Another initiative may be to ask a class to work together to print an anatomical model that would otherwise cost Baptist College thousands of dollars to purchase.

Figure 5.5. Twitter poster for a new print. *Nichelle Mack.*

NEW HORIZONS IN COLLECTION DEVELOPMENT: 3D PRINTING AND MODEL CREATION

Dorothy Ogdon, MSIS, AHIP
Lister Hill Library of Health Sciences
University of Alabama at Birmingham

In early 2017, a student at the University of Alabama at Birmingham (UAB) contacted the Lister Hill Library of the Health Sciences to inquire about the availability of a human skeleton to use to study for an academic examination. Though the library has provided access to this type of model previously, at the time of the student's request, there were no models available for loan. Though health sciences libraries are often interested in providing access to anatomical and molecular models to support student learning, the cost of these models is often prohibitive, and the development and management of this type of collection is often not sustainable due to financial limitations. Following the student's request and discussions of this topic in the library's reference department, Dorothy Ogdon, reference librarian and liaison to the UAB School of Dentistry, approached Kay Hogan Smith, senior research librarian and liaison to the UAB School of Public Health, to discuss the potential use of 3D printers to create models for use in the library. Smith is chair of the UAB Libraries Grants Committee and was interested in bringing 3D-printing technologies to Lister Hill Library. Ogdon subsequently wrote and submitted a project proposal, "A Novel Use of 3D Printing Technologies for Library Collection Development," to the University of Alabama at Birmingham's Faculty Development Program (University of Alabama at Birmingham Faculty, n.d.). She proposed investigating whether 3D printing could be used to create anatomical and molecular models of a high enough quality to facilitate student learning. Source files for the models were accessed through the NIH 3D Print Exchange, a newly developed resource provided by the National Institutes of Health (US Department of Health and Human Services, National Institutes of Health, n.d.). In addition to creating models using 3D-printing technology, Ogdon aimed make the models available to the UAB campus community as a part of a circulating collection in Lister Hill Library. Adding the models to a library collection creates a new shared resource for the campus community. Students, faculty, and staff are able use models as needed at no cost rather than having to negotiate a new expense to further research and instruction.

The proposed project was submitted to the Faculty Development Program in February 2017 and was selected for funding in May 2017. The final outcomes of the project were presented on November 14, 2018, during the Faculty Development Grant Program poster session. At the time of the poster session, there were a total of twenty discrete objects in the collection (University of Alabama at Birmingham Faculty, n.d.), including models of the human heart, human brain, human vertebrae, and the skull of Phineas Gage (see figure 5.7). As of November 4, 2018, the models have been checked out a total of thirty-two times. The project has also generated several objects that were not added to the library's collection due to print quality issues, including warping, shifting, stringing, and, in one case, filament supply running out before a print was completed. The models that have not been added to the library's collections are being used to provide information about 3D printing and object creation during lectures and other scholarly activities.

Requests for models have been collected from students and faculty through informal interactions; through the library's 3D-printing guide (University of Alabama at Birmingham Libraries, 2018); and during library tea, a weekly library marketing event held Wednesdays at noon during the fall and spring academic terms. The models most frequently requested during library tea include a model of a pediatric skeleton, as well as models of the kidney and liver. Students have also requested different visualizations of organs and their positions within the body and have often used photographs and illustrations from print anatomy resources as examples.

This project has increased awareness of 3D-printing technology, and faculty members have expressed interest in developing 3D-printed models to enhance instruction; there is also interest in exploring potential applications of 3D printing in simulation. The UAB libraries are preparing to launch a mediated 3D printing space in Lister Hill Library in Spring 2019. The space will include two LulzBot Taz 6 3D printers, a Form 2 stereolithographic printer, and a FlashForge Finder printer. Ogdon will continue her involvement in 3D printing in the UAB libraries by leading the development of the new 3D printing space; delivering related instruction; providing collection-development services; and investigating other approaches to support faculty, staff, and student use of 3D printing.

Innovative 3D printing and makerspace-related activities at health sciences libraries in the United States have heavily influenced both the development of the initial project proposal and subsequent ideas for library programming and academic applications of 3D printing. Jennifer Herron's presentation at the 2017 meeting of the Medical Library Association provided initial inspiration for promoting 3D printing among health sciences students (Herron, 2017). The March 6, 2018, HS/HSL Maker Expo at the University of Maryland Health Sciences and Human Services Library in Baltimore, Maryland, provided insights into the maker health movement and related research and clinical projects (MakerHealth, 2016; Zelip, 2018). The free workshop following the expo, presented by Jeremy Swan, included the use of Blender software and the Protein Data Bank to create three-dimensional molecular models, an application of 3D printing that is anticipated to be in high demand as interest in this technology grows within in the UAB community (RCSB Protein Data Bank, n.d.; Swan, 2015). The HS/HSL Innovation Space and the work of Brian Zelip have also been a crucial source of ideas and information that have been used to develop the 3D printing space in Lister Hill Library (University of Maryland Health Sciences and Human Services Library, 2018a, 2018b). Finally, the 3D printing service developed by Richard Sexton of the Pendergrass Agriculture and Veterinary Medicine Library at the University of Tennessee in Knoxville has been an extremely useful example of both projects and processes to facilitate 3D printing as a functional and valued service in an academic health sciences library (University of Tennessee, Knoxville, 2018).

Because 3D printing has only recently become widely available and accessible, the understanding of potential applications for this technology as a tool to support student learning, research, and even clinical services is still limited. There are a number of applications for this technology in academic libraries that have yet to be explored, including opportunities to develop new collections, services, and areas of technical expertise within the library profession.

3D PRINTING AT THE UNIVERSITY OF ARIZONA HEALTH SCIENCES LIBRARY

Michelle Halla, MLIS
Yamila El-Khayat, MA
University of Arizona Health Sciences Library

The University of Arizona (UA) Health Sciences Library (HSL) implemented a successful 3D printing service in March 2016. 3D printing as a service had already been in place for a year at the UA Libraries' Main Library and was expanded to HSL after merging access services with the University of Arizona Libraries system.

About the Library

The Health Sciences Library serves the UA Colleges of Medicine, Nursing, and Pharmacy and the Mel and Enid Zuckerman College of Public Health. The library is physically connected to a hospital and also serves medical professionals, as well as community members doing medical or health-related research. While originally a separate entity on campus, the library became part of the campus library system, University of Arizona Libraries, in 2015 and includes the Main Library, Science-Engineering Library, Fine Arts Library, and Special Collections.

Equipment and Supplies

When 3D printing first launched at HSL, the service used MakerBot Replicator 2s. These printers are easy to use (which is important because they are mostly operated by our student employees, who have varying comfort levels with technology); moderately easy to troubleshoot and repair; and sturdy. They use a glass build plate that requires a sheet of blue-painter's-tape-like material to be placed on top to increase print adhesions. The filament used with these printers is PLA, a plant-based biodegradable plastic. Like many similar consumer models, these printers are prone to clogging when operating at the high volume of a printing service.

In December 2016, the libraries refreshed all printers at both the Main Library and the Health Sciences Library. The new printers purchased were Zortrax's M200 model. These printers print in ABS and feature high resolution and a heated, perforated build plate. This material is stronger than PLA but less environmentally friendly and requires ventilation. While the resolution of the M200 is impressive, especially when printing anatomical models, the ABS material is more prone to warping than PLA.

Other supplies purchased included spatulas to remove objects from the print bed; a set of tools, including hex keys and screwdrivers in multiple sizes; and a small scale.

Service Model

Our service is open to both UA affiliates (students, faculty, and staff) and community members. When expanding this service, the 3D printing leads had to determine whether we would begin limiting customers to specific libraries. Students from the health sciences were already printing at Main Library; did we want to require that they print at HSL, and likewise, did we want to restrict Health Sciences Library printing to just health sciences students? The 3D printing leads already knew that some non-health-sciences students prefer studying at HSL, either due to location or ambiance, and decided to have the option for students, faculty, and staff in any department submit to any location.

Our pricing model charges only for the cost of materials used, while the libraries subsidize the labor. When our service began, this translated to charging ten cents per gram of material of the final product (not including any failed prints), and prints under one dollar were free. Our integrated library system (Millennium) didn't allow us to add charges under one dollar to accounts, and the cost associated with not charging for these prints was overall low. Because many commercial pay-to-print services must charge for both labor and supplies, as well as turn a profit, they are about five times as expensive for the same objects and materials.

In April 2018, 3D printing leads changed the pricing model to include tiered charges. This was to reflect changes in the service, such as offering more types of filament, and the increase in highly complex objects that required significant staff time to evaluate and set up. Basic printing is still 10 cents per gram, but minimum charges are $1.01, regardless of amount of materials used. Printing with specialty filament (such as higher-grade materials or materials with special properties like flexible or magnetic filament) is now fifteen cents per gram. Complex jobs that require additional staff time and effort have a two-dollar set-up fee.

Workflow

There are a lot of benefits to expanding an existing service over implementing a brand-new service, such as using the same brand of printer so that supplies, like filament or back-up parts, can be shared between locations. However, there are some different types of challenges, such as deciding to modify existing workflows or create new workflows to accommodate the growing service and deciding to keep processes alike for all locations or modify based on unique needs of one service site.

First, let's discuss the process as it appears to our users. Perhaps they found something from a website, like Thingiverse, where people shared 3D files for free, or perhaps they used computer-aided design (CAD) software to create their own objects. To print an object in 3D, the user completes a web form and uploads the object as a stereolithography (.stl) file. They can choose to have an estimate with cost and turnaround time e-mailed to them before approving it to be printed. The customer is notified when the item is completed. UA affiliates have charges added to their library accounts, which they can pay in person or online or by using a department account. Community users, however, don't necessarily have a borrower's card and accompanying library account. Instead, they come to the service desk to pick up their receipts and pay in person, and then their objects are released to them.

Second, let's review what the process looks like on the back end. The web form is supported through Springshare's AskUs product, as the libraries were already using this product to manage our e-mail reference. This provides a way for 3D printing staff to correspond with customers from a designated 3D printing e-mail address, 3D@lib.arizona.edu, and allows multiple staff to view tickets. Staff download all new submissions and enter the information for the requests into a "Master Jobs" spreadsheet. Individual jobs are evaluated for printability, such as the orientation and needed support structures. Often items will need to be resized if the submission is too large to fit on the print bed or too small to print properly (usually due to the software converting measurements in inches to millimeters). If the customer requests an estimate, that is e-mailed. Once the response is received, the item is added to the "Queue," a separate spreadsheet, and our student employees operate the 3D printers to complete the jobs. After the object is completed, it is weighed, and charges are added to a customer's library account. The customer is contacted to let them know that the object is ready and they must pay for the item before picking it up. At the desk, staff check customers' accounts to ensure prints have been paid for before releasing items.

One of the first decisions the 3D printing leads needed to make was whether to create a separate web form and e-mail address for HSL printing. Because the same staff members would be processing prints and corresponding with customers regardless of pick-up location, we opted to keep the single web form and e-mail address. We determined this would be less confusing for customers and staff alike. However, we already had a year's worth of data in our old "Master Jobs" spreadsheet, which used a lot of internal formulas for data collection. We needed to be able to add additional fields (such as printing location: Main or HSL) without breaking the spreadsheet.

Another hurdle to cross when expanding this service to the Health Sciences Library was how we wanted to direct community, or non-UA, customers. As mentioned previously, community users of the service paid for their items in person. At the time, the Main Library had a desk that processed payments, but the Health Sciences Library did not. The 3D printing leads considered directing all community users to print at the Main Library, but we also wanted to have 3D printing services available to hospital employees who were not considered affiliates but worked in the adjacent hospital. We compromised by creating user accounts for hospital employees that would allow them to pay for their prints online and directed all other types of community use to the Main Library.

By the Numbers

Collecting data has been an important part of our service from the beginning. In addition to the information required to successfully print an object, our web form asks users about their affiliation, their department, whether their object is an original design, and whether the submission is for a class. We collect this data for all submissions and can look at it holistically or by pick-up location.

Over the course of offering 3D printing at the libraries, this robust data has served us in many ways. 3D printing staff can confirm that we've had submissions from nearly all colleges on campus (not just engineering students want to print in 3D!). 3D printing leads provide our liaisons with the names of instructors who have used the service in their curriculums so that liaisons can connect and offer additional library support and resources. This robust data also allowed the 3D printing leads to identify trends that are the same for both locations, as well as identify the ways in which the two pick-up locations differ.

For example, 3D-printing submissions dropped 42 percent in the second year at HSL. During the same time period, the number of original designs submitted nearly doubled. 3D printing leads saw similar but less drastic trends at the Main Library. One interpretation of this data is that there is an original peak of interest, where people are introduced to the idea of 3D printing, and many want to try it out. Out of this group, some are inspired to use 3D printing to create their own tools, make gifts for their friends, or complete class projects. While high submissions are impressive, I am more impressed when more than half of submissions are students' original designs (see figure 5.6).

Year	Total HSL jobs submitted	Total original designs	% of total HSL submissions
2016 (March – Dec)	402	66	16%
2017	235	127	54.04%
2018 (Jan – May)	50	26	52%

Figure 5.6. Summary of 3D Print Requests.

User Story

We frequently reach out to users who submit original designs to find out more information and request permission to use their items for outreach and marketing purposes. User stories are employed to inspire other students; to communicate the value of the service to library administration and student governance organizations; during donor visits, to demonstrate ways technology is incorporated into curricula; and for other general promotional and marketing materials. A broad range of examples ensures that you have something that appeals to people from every discipline. Examples of how students are using a technology tends to be more impressive and more reachable than some of the abstract applications we may see in the general news. User stories help make advanced technology seem approachable.

One of my favorite examples of how 3D printing is used in a medical library comes from student Davis McGregor. McGregor was working in the Tissue Optics Lab, an interdisciplinary campus that aims to "improve healthcare through the novel use of light," and student work includes "building new imaging devices, testing new applications, and performing signal and image analysis" (University of Arizona Tissue Optics Lab, n.d.). McGregor was working on a project that used multiple types of imaging on mouse ovaries to search for precancerous tissue abnormalities. McGregor created an "ovary spoon," a device whose purpose was "to hold the mouse ovary outside the body during in vivo imaging . . . the tissue need[ed] to remain stationary during the long image acquisition, thus isolating from the breathing mouse body is essential" (McGregor, 2016).

In a process known as iterative design, McGregor submitted a few different models, tested them in the lab, and tweaked the design several times before finalizing it. The spoons were relatively small and printed quickly, which allowed him fast turnaround time between each design iteration and real-life testing. The low cost of 3D printing meant that, when finalized, McGregor could print these in bulk and then discard them after use:

> The overall goal of the research project is to search for precancerous tissue abnormalities. Extrapolating this data to humans, this would allow us to use an endoscope to scan the ovaries and fallopian tubes of high-risk women for any

> precancerous indicators and hopefully reduce the number of preventive hysterectomies in young women. (McGregor, 2016)

3D Printing and Outreach

We have also started to experiment with taking 3D printers outside the library space and into the communities we serve. As a land-grant institution, the University of Arizona has a commitment to serve all communities in the state of Arizona. Our HSL has a position specifically funded to do outreach within the state in an effort to increase health literacy. In 2016, our librarian suggested taking a 3D printer with her to a health fair. The response was impressive. The numbers of people she reached more than doubled. She noticed that having this type of technology was the key to attracting people to learn more about our libraries and the services we have to offer. Once people came to the table, she would talk to them about the 3D printing service, show examples of work done with it, and then talk to them about open-source, health-related databases to learn more about health.

Many of the rural communities that our outreach program reaches do not have access to these new technologies, and 3D printing remains a thing of the future. Having the opportunities for these communities to witness a 3D printer in action and ask questions has proven to be a valuable experience, as many express their gratitude for having the chance to see a printer like these. Our outreach librarian has been invited to host technology showcases in these communities, with possibilities of developing summer camps for youth interested in medicine. These camps would introduce them to some of the projects that others have done here at HSL and have them experiment on their own, delving into their creativity to transform health care with these technologies.

The number of people reached using 3D printing in outreach has grown exponentially. When comparing data from previous years, our larger health fairs reached somewhere between three hundred and four hundred people. With a 3D-printer exhibit at a table, we now reach closer to eight hundred. Our programming is quickly growing, as requests constantly come in for programming that uses these new technologies, especially programming for children in K–12 schools.

Conclusion

HSL has been very successful in its implementation of this new service. As noted earlier, the number of print jobs at the Health Sciences Library has declined, but the amount of original work has grown. This may be because people are now digging into their own curiosity to create tools and products that can be used in their environments and that fit all the specifications they require. In outreach, the service has also been a true success; evidence for this is the amount of people now being reached along with the number of outreach requests coming into the library. As a 3D printing provider, we are constantly looking for new ways to improve our service and are now exploring the possibility of refreshing our printers to newer equipment every year.

RUTH LILLY MEDICAL LIBRARY—THE NEXUS MAKERSPACE 3D PRINT LAB

Jennifer Herron, MLIS, former emerging technologies librarian
Ruth Lilly Medical Library, Indiana University School of Medicine

The Ruth Lilly Medical Library (RLML) serves as the main library for Indiana University School of Medicine. The school has nine campuses throughout the state of Indiana that reach as far north as South Bend and as far south as Evansville. These two campuses are roughly two and a half hours from Indianapolis, where RLML is located. RLML began its 3D printing service in summer 2015, and the service is still running strongly to this day, with requests becoming more advanced the more familiar with the technology that users become.

The Printers

With its new service, the library started off with two Cube printers: a Cube and a Cube Pro Trio. The Cube can print smaller models, with a print bed of six inches by six inches by six inches, and has two extruders for two filaments to be used. This can be either multiple colors or PLA filament with rinse-away material for supports. One drawback with using multiple extruders is that more time is needed to complete prints. The larger printer, the Cube Pro Trio, can print much larger models, up to the size of a basketball. Additionally, the Trio can print three different materials at the same time, allowing for either multimaterial or multicolor models to be printed. The two printers were later joined by a MakerBot Replicator 2, given to the library by the University Information Technology Services (UITS). In the third year, an Ultimaker 3+ was added. In addition to the printers themselves, the library also purchased two scanners to assist in modeling requests. One scanner, Cubify Sense, attaches to a laptop and is able to scan large objects, including successfully scanning two staff members. The other scanner, an iSense, attaches to an iPad for more mobility, but the trade-off is that only smaller models can be scanned.

Getting Started

Not having any experience in 3D printing, the library soft-launched the service to better acquaint themselves with the technology and not become overwhelmed by requests without completely understanding the printing process. The first request for 3D printing was a great test of librarians' 3D printing knowledge and the capabilities of the library's 3D printers. Not only did the printers fail to print the request, a model of a neuron, but librarians also failed to recognize that the model was too intricate for the printers. The student who requested the print was very understanding and submitted a new request that was successfully printed using the Cube Trio with all three extruders for a multicolor print with rinse-away support. Having identified the knowledge gaps in our service, our trio of librarians operating the 3D printing service decided to take trips to visit the other 3D printing labs on campus.

Field Trips

After some investigating, it was learned that there were three other 3D-printing labs on campus. The university library on campus has a small print lab managed by UITS. UITS also opened a larger print lab housed in the university's informatics building to support the increased demand from university students, staff, and faculty. The smaller library print lab houses a single MakerBot and has two student employees who assist in design and printing. The UITS lab has multiple MakerBots and a large-scale printer, the MakerBot Z18. This lab also supports design and postprocessing efforts. The lab works with students and faculty across the university and collaborates with another UITS lab located in a library on Indiana University's Bloomington campus.

The UITS lab referred us to another niche 3D print lab located in the Herron School of Art's main building. The Think It Make It (TIMI) lab worked with art students, as well as other university students, and offers FormLab resin-based printers. Additionally, students are able to use the printers and have access to software installed on computers in their lab. The lab has plenty of room for collaboration and workstations. A laser engraver is also available for free use.

A trip to Bloomington allowed our library team to also visit the other UITS print lab. Similar to the Indianapolis campus, the Bloomington UITS lab is located in the campus library. The lab has MakerBot printers, provides students with design support, and collaborates with such outside organizations as local schools to promote 3D printing as a learning tool.

All of the labs visited provided the RLML print lab with good insights as to model printability, printer capabilities, postprocessing, and even how to deliver models. Furthermore, UITS offered its support to troubleshoot printer problems and welcomed the chance to collaborate with original model designs from faculty members.

Printing Progress

As the library became more proficient in 3D printing and students saw each others' completed models, requests increased. The library supports students at all nine IU campuses, and as such, timeliness is important; shipping adds a delay in the student receiving the model. In the first year, only a few print requests were received, with anatomy models being the most popular. The second year saw a huge increase in requests, with more than one hundred requests from multiple campuses. Anatomy models continued to be a popular request, but faculty and staff also took advantage of the printers for testing their own designs, printing parts for creative endeavors, and creating proof-of-concept tests. The library offers complete postprocessing support and even paints models on request (e.g., a periwinkle brain and a pink skull). The third year saw somewhat of a decline in the number of print requests; however, the complexity of prints increased significantly. Students were no longer asking for simple skulls or brains; they now were requesting complete lower limbs, listing the individual bones wanted; pathology models; complete skeletons; and hearts sectioned off to show interior anatomy. Luckily, the library hired two student assistants during the second year of printing who helped maintain the printers as well as worked with faculty and students to design models. Our student employees also helped to clean up and edit models to ensure printability.

During the first year of operating the print service, the library helped to organize a 3D printing group that pulled together the 3D print labs and other individuals with 3D printers. Students in informatics, as well as faculty, also joined to collaborate on projects. Through this group, the library made connections with engineering, informatics, and the campus's Advanced Visualization Lab (AVL). Additionally, a staff member from the Indiana School for the Blind and Visually Impaired joined the group after collaborating and working with faculty from health and rehabilitation. The AVL shared their expertise on generating printable models from Digital Imaging and Communications in Medicine (DICOM) and also looped the library in on some medical 3D-printing projects that they had previously worked on. The AVL provided the library with training on using Slicer and shared their workflows. The informatics members helped to troubleshoot design issues, and one of their members also taught a course on medical 3D printing. They similarly informed the library about projects they worked on with medical school faculty and increased awareness about the library's 3D printing service.

Challenges in 3D Printing

Some of the challenges over the years as the library developed its 3D printing service include training and maintenance of the printers. One of the printers, the Cube Pro Trio, was inactive for almost a year due to maintenance issues. The time requirement for maintaining the printers was extensive at times, slowed the service down, and caused delays for delivering models. Other challenges included design issues and troubleshooting design problems that caused models to fail. Open-source software was found, which offered model inspection and repair, but this did not always correct some issues, or it created new issues that needed to be repaired. Hiring a student assistant with a background in 3D model design and software greatly helped, increasing efficiency and reducing the number of failed prints.

A 3D Printing Event

During the second year of the 3D printing service, the library decided to host a MakerHealth Faire, partnering with other departments on campus who were in the 3D printing group. Along with the faire, a panel of faculty and staff in health and medicine who use 3D printing presented on their work. The event was held in the spring 2017 and had speakers from engineering, health and rehabilitation, dentistry, informatics, and the Advanced Visualization Lab. Tables were hosted by UITS; the AVL; School of Health and Rehabilitation; Mechanical Engineering; and a community makerspace, First MakerSpace; as well as staff from the Indiana School for the Blind and Visually Impaired. Additionally, the library invited a local makerspace to host a table and promote design thinking. The event created an opportunity for faculty and staff from a variety of disciplines to get together and collaborate. Furthermore, it increased the library's presence with 3D printing on campus. The library was able to learn about new 3D printing initiatives on campus, and the library was able to offer its insights and recommendations as needed. The collaboration between the library and other 3D printing services on campus increased and provided the library with its own homegrown team to help support the printers and troubleshoot fails.

Conclusion

Though I am no longer at the Ruth Lilly Medical Library, it was a great experience to be a part of the new 3D printing service. It continues to draw interest, and in April 2018, a new event was held to showcase 3D printing. The Health Technology Symposium offered a chance to promote 3D printing, and again, experts were invited to share their stories and experience with the service. As the technology grows and people become more aware of its capabilities, the interest will grow as well, and the library will be the perfect place for 3D printing to be offered to support users' interests, education, and professional endeavors.

Memorable prints:

- Phineas Gage skull, to demonstrate case history (figure 5.7)
- EEG headset, for research purposes (figure 5.8)
- Brachial plexus, for anatomy studying (figure 5.9)

Sadly, not pictured but worth noting:

- Pangolin, stress-reliever request and a test of the printer's capabilities
- T-Rex skeleton, requested for comparable anatomy
- Body-part necklace, an amazing marketing idea from the outreach librarian

REFERENCES

Herron, J. 2017. "Finding a Niche: Where 3D Printing Services Can Have the Most Impact." Medical Library Association meeting May 29, 2017, Seattle, WA.

MakerHealth. 2016. http://www.makerhealth.co.

McGregor, D. (2016, November 28). Email communication with M. Halla.

RCSB Protein Data Bank. n.d. "A Structural View of Biology." Accessed June 30, 2018. https://www.rcsb.org.

Swan, J. 2015. "Home." https://science.nichd.nih.gov/confluence/display/~swanjere/Home.

University of Alabama at Birmingham Faculty. n.d. "Faculty Senate." Accessed June 29, 2018. http://www.uab.edu/faculty/senate/140-faculty-development-grant-program.

University of Alabama at Birmingham Libraries. 2018. "3D Printers at Lister Hill Library: Home." https://guides.library.uab.edu/LHL3DP.

University of Arizona Tissue Optics Lab. n.d. "Welcome to the Tissue Optics Lab!" Accessed September 25, 2018. http://bmeoptics.engr.arizona.edu.

University of Maryland Health Sciences and Human Services Library. 2018a. "HS/HSL Innovation Space." https://www.hshsl.umaryland.edu/services/ispace.

———. 2018b. "Staff Profile: Brian M. Zelip, MSLIS, MA." https://www.hshsl.umaryland.edu/about/staff/profile.cfm?id=bzelip.

University of Tennessee, Knoxville. 2018. "3D Printing @ Pendergrass." https://libguides.utk.edu/c.php?g=188821&p=1244830.

US Department of Health and Human Services, National Institutes of Health. n.d. "NIH 3D Print Exchange." Accessed June 30, 2018. https://3dprint.nih.gov.

Zelip, B. 2018. "HS/HSL Maker Expo." University of Maryland Health Sciences and Human Services Library. https://www2.hshsl.umaryland.edu/hslupdates/?p=3095.

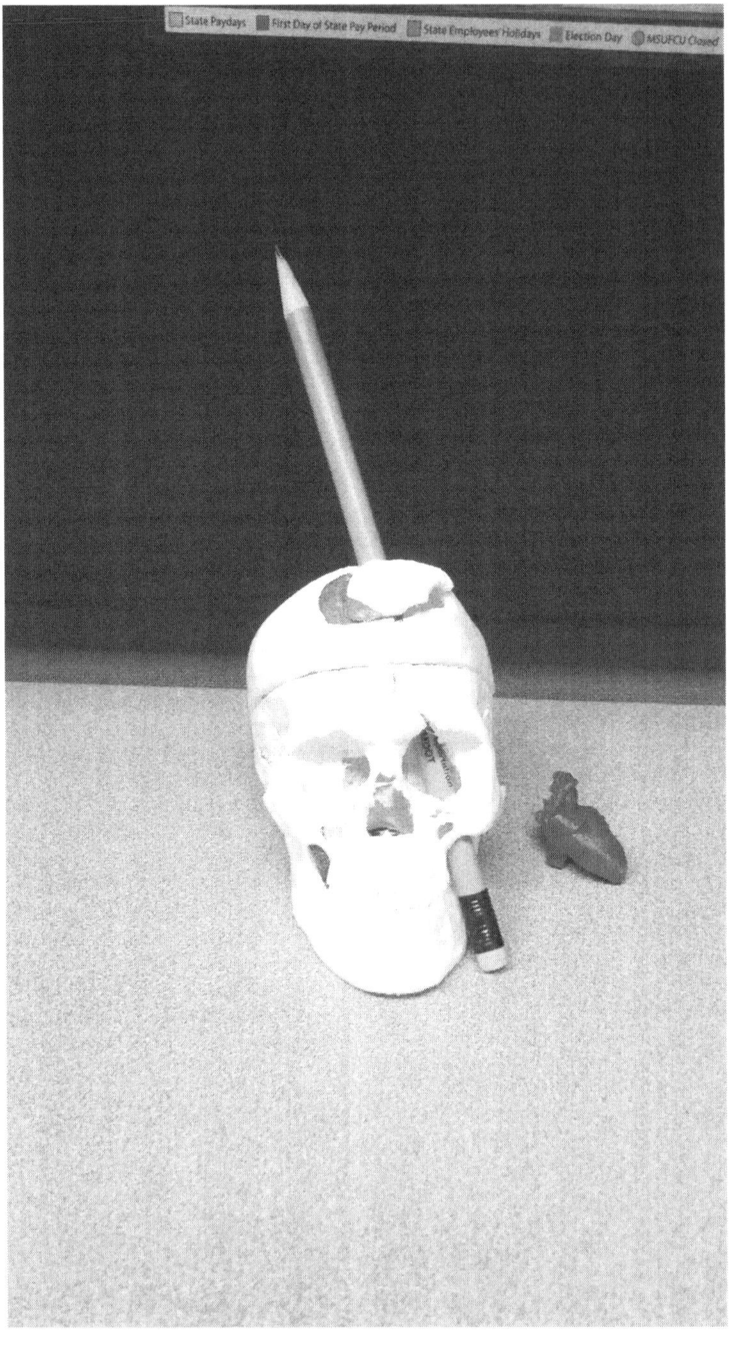

Figure 5.7. Phineas Gage skull at 50 percent. ***Jennifer Herron.***

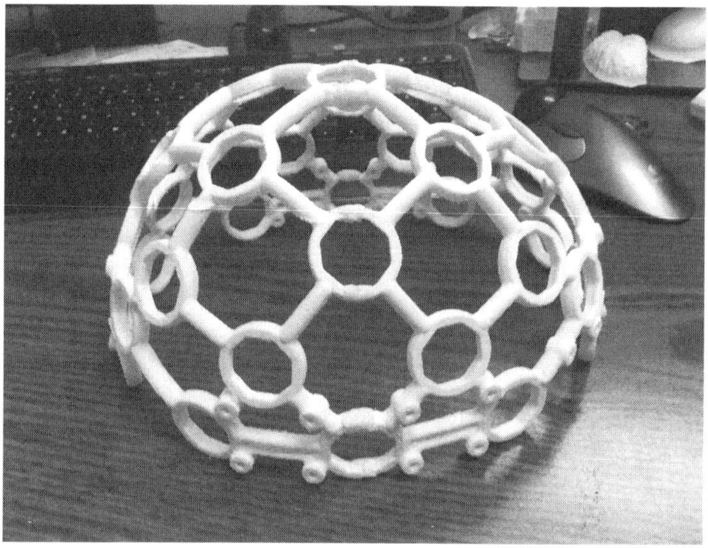

Figure 5.8. EEG headset for researchers. ***Jennifer Herron.***

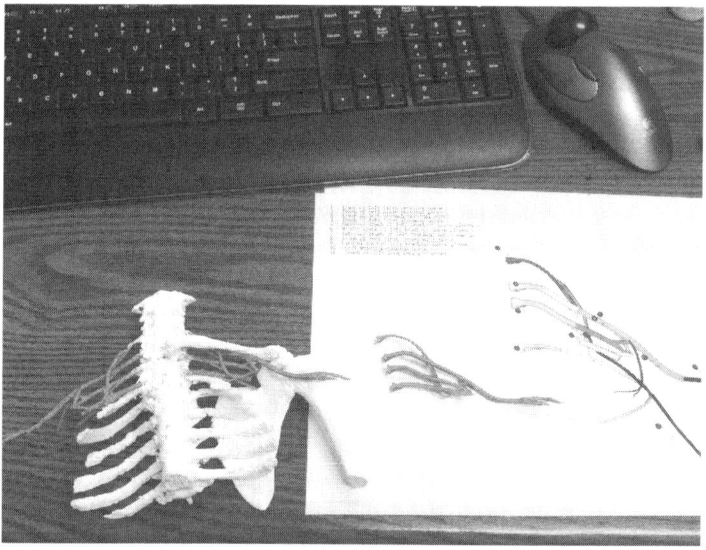

Figure 5.9. Brachial plexus model made using BodyParts 3D. ***Jennifer Herron.***

Chapter Six

Finding 3D Models for Anatomy

Depending on how your 3D printing service is operated, you may require students to bring their model files to print, or they may request a search for models. Searching for models is very much like any other information search and requires the same methods and techniques to determine quality of the sources provided. One technique for evaluating sources is the CRAAP method (currency, relevancy, authority, accuracy, and purpose; Blakeslee, 2004).

When you treat virtual objects like any other information source, you will find different techniques and indicators to determine quality. The following is a breakdown of the CRAAP method and how to apply it to 3D-printable models and sources.

CURRENCY

This is not as much of an issue with 3D models, but instead of currency, think of versions. 3D models that have been uploaded to a repository can undergo revisions, and you might find a version of a model with design flaws that were corrected in another version. Similarly, because of the open and collaborative nature of 3D printing, models may have been altered from the original to either incorporate new elements or for users to personalize.

RELEVANCY

Some of the resources discussed later in this chapter have a very specific audience in mind: other health professionals. Typically, relevant information in included on the main resource information or "About" page. If it is not clear, the model page itself usually provides a background on the origin and conveys any other information about the model that requestors might need.

AUTHORITY

Determining the creator of a model is usually not too difficult; however, determining the creator's background and credentials may prove challenging in some resources. This is when you need to work with your requestor and pull from their experience to provide authority to a model. It is important to explain the model's source to requestors so they know potential credibility issues, where their model came from, and potential inaccuracies that might exist. Examples of these situations can occur when searching a nonmedical or health-focused resource where the creators do not share their backgrounds or processes they used to create the model. One way to alleviate this issue when searching for models is to e-mail the requestors links to websites of potential models and have them make the final decision on which model to print. In this way, you are able to include requestors more in the 3D printing process and allow them to share the responsibility of determining a model's credibility. Furthermore, after the model has been printed, you can ask them for their opinions on whether the model fulfills their needs and if others might benefit from it. This helps to vet models in-house and creates your own set of go-to models for printing.

ACCURACY

Trying to review the accuracy of a model is extremely difficult as librarians are not expected to be anatomists, chemists, or lab-ware specialists. By establishing the model's relevancy and authority, the accuracy can be inferred, but again, this is an area where involving the requestor is important. Furthermore, problems may arise during printing that cause some inaccuracies in the final model. It is recommended that these potential problems be clearly explained in the 3D printing service policy.

PURPOSE

Models shared online are typically shared for one reason: to inform and educate. Models are shared to educate other health professionals about anatomy and pathology. Additionally, models might also be shared as a platform to collaborate with others to improve an idea.

When searching for models, it is important to know who created the model to get a better understanding of that person's background and their authority on the subject matter. This is one of the most challenging aspects of

3D printing because not all model creators are subject-matter experts and not all subject-matter experts have the ability to create models. New software has greatly diminished this issue, but reviewing models is still necessary to ensure credibility.

Librarians' expertise in evaluating information and sorting through irrelevant and less-than-credible resources is a great value for a 3D printing service. It makes the health sciences and medical 3D printing labs unique and sets them apart from academic and private print labs. In addition to aiding students and faculty, librarian searches boost the 3D printing service by saving time searching for models, helping to created models that may never have been printed due to the effort involved in searching. One of the amazing things to see with 3D printing in health and medicine is the open-source sentiment that seems to be getting larger every day. Some resources may start to charge for select or higher-quality models, as is the case with Embodi3D. However, with more people developing the skills to create models from scratch and from DICOM data, there likely will be an influx of model availability at no cost.

UNDERSTANDING MODEL TYPES

One thing to determine before you begin searching for models is the model's purpose. What does the requestor want to do with the finished model? Do they want to use it as a study aid? Is this more of an exploratory request to see what is possible with 3D printing? Are they interested in normal anatomy or abnormal anatomy? When a request is received, it should be treated just like a normal reference interview, and follow-up questions should be asked to make sure that the search is timely and effective. In my experience with 3D printing, I found that there are five main types of model requests:

1. **Normal Anatomy:** Models typically used as study aids and sometimes for "inspiration"
2. **Pathology:** Models used to learn more about abnormal anatomy
3. **Molecular and Biochemistry:** Models used to better understand molecular biology and biochemistry
4. **Lab Support:** Models used to help with lab work and may involve some prototyping
5. **Stress Relievers:** Anything fun that can be deemed "stress relief" for students (if your 3D print service allows for nonmedical or noneducational prints)

By understanding the type of model being requested, different resources can be used to speed up the search. For busy print labs with numerous requests, the faster a print can be started, the faster the queue can move. Again, your 3D print service filling requests for models means not only finding a model but also prepping the model for printing. More about this process is discussed in later chapters.

READY-REFERENCE SOURCES

In traditional information searches, there are ready-reference resources where librarians can turn to for quick and reliable answers. The same is true when searching for 3D models. The following is a list of some go-to resources for health and medical 3D models. Additional information includes best bets for the types of models you can expect to find and the ease of searching and locating models.

BodyParts3D
Difficulty: hard
Best Bet: normal anatomy

Embodi3D
Difficulty: medium
Best Bet: normal anatomy, pathology

Google
Difficulty: expert (Google is set at its own level because of the complexity involved in the searching process. You will need to use a variety of search strategies to retrieve relevant results, and even after this, it will be time-consuming to review individual results.)
Best Bet: normal anatomy, pathology, molecular and biochemistry, lab support

NIH 3D Print Exchange
Difficulty: easy
Best Bet: pathology, molecular and biochemistry, lab support

Thingiverse
Difficulty: easy to medium
Best Bet: normal anatomy, pathology, molecular and biochemistry, lab support

RESOURCE BREAKDOWN

BodyParts3D (http://lifesciencedb.jp/bp3d/)

Background

Created in 2007, BodyParts3D is a self-described "anatomical dictionary" created by the Database Center for Life Sciences of the Japanese Research Institute. Models were created from a male volunteer who had his entire body scanned. The website is based out of Japan and is in Japanese, but translation is available on the site itself, with an "English" button in the upper-right corner; web browsers should also offer this service when the website loads.

What's Included

Anatomical models are included and segmented by "Bone," "Muscle," "Vessel," "Internal," and "All." Furthermore, models are sectioned off by head and neck, midbody, and lower limbs. The website lists 68,440 material anatomical entities that were created using medical scans and cleaned up yet maintained anatomical correctness using artistic rendering. Anatomical segments are able to be rendered and viewed as a virtual 3D model using the "Anatomography" feature, and users can download an image with color-coding, as well as the 3D printable file (Mitsuhashi, 2009).

Evaluation

Currency: Data versions do not include explicit date information, but the models are not likely to change in content or context.

Relevance: The intended audience is also not explicit; however, the complexities included in the anatomy and the detailed methods of searching and browsing lead to the service being primarily useful for anatomists and those studying anatomy.

Authority and Accuracy: Models were created using real medical scans from an MRI and any gaps were supplemented with medical illustrators helping to complete designs. The organization, which created BodyParts3D, is cleared and listed and provides good authority over the accuracy of the content.

Purpose: Given that no customization is allowed and models include only healthy or normal anatomy structures, the purpose of this resource is primarily informational as a supplement to anatomy education.

Search Features: Searches can be completed using the search box or browsing the segmented sections. The data used for the model files is defaulted to the heart. To browse outside the heart, there is a drop-down menu to change the data version. There is also a data set for the brain, where more detailed model files can be searched.

Model Details: Models are focused on normal anatomy and would support students seeking more-detailed anatomical models and those interested in internal anatomy.

Embodi3D (https://www.embodi3d.com)

Background

Embodi3D was created by interventional radiologist Dr. Michael Itagaki when he learned how 3D printing could affect patient care after he used it for a difficult case (Hanna, 2016). When he found himself without anyone to consult about the applications he was considering for his patient, he realized that a community was needed, where health experts could collaborate and help each other use 3D-printing technology. The mission of Embodi3D is "to democratize medical 3D printing by making it easily accessible to anyone, through tutorials, blogs, discussion forums, a marketplace, and software" (Embodi3D, n.d.).

What's Included

Embodi3D includes anatomical models created from medical scans. Anatomy models are divided into six categories: bones, muscles, cardiac and vascular, organs of the body, skin, and veterinary. Categories are further divided into more-specific regions of anatomy. Additional categories include science and research, miscellaneous, and medical CT scan files.

Embodi3D also offers a free program, Democratiz3D, for converting DICOM files into printable models. For users needing a higher-quality resolution for their generated models, a premium subscription is available. Models that are processed with the higher resolution can either be downloaded by users with a premium membership or are available at a cost to download.

Evaluation

Currency: A versioning system is in place; models and updates are clearly listed on pages with links to newer or previous versions. Model pages also include tracking information for the latest update. A "Comments" section also exists for users to share notes about the model and potential printing or design issues.

Relevance: Embodi3D is noted as a "Biomedical 3D Printing Community" and, as such, indicates its intended audience.

Authority and Accuracy: The user community, while not exclusive, is mainly health professionals. Models are rated and reviewed by other members in the community. This is not a requirement but allows for accuracy to be disputed if the need arises. Member profiles also can provide some information on the users' backgrounds and includes a "Rank and Reputation" section to learn more about how they've interacted with the community.

Purpose: The main objective of Embodi3D is to create and share scans and models of real patient anatomy. Users can decide what they do with these models, and examples of use are listed on Embodi3D's "About" page. These include to aid in creating medical devices, as illustrative tools to demonstrate patient injuries for legal cases or other educational purposes, and for training.

Search Features: Users can browse Embodi3D using their categories, and it does provide a search option. Additional search options allow users to select files for content type and filter out any nonmodel results (blogs, images, etc.).

Model Details: Models are focused on both normal anatomy and pathology, with a concentration on bones (more than 1,200 as of June 2018). Muscles are the next-popular category, with 139 models.

Google (http://www.google.com)

There are many potential resources that are uncovered in Google results. These need to be reviewed and may lead to many wild-goose chases but may also lead to the discovery of new and great resources. See the chapter 14 for resources discovered using Google.

Background

Google is a well-known search engine and a quick go-to for casual searchers.

What's Included

Advanced search options can provide some ability to make a refined search, but it is not perfect and requires that results are carefully reviewed.

Evaluation

Currency: Models found using Google are most likely found in other, larger repositories of 3D models. These include those that must be purchased, those that are used for virtual reality and digital models without the ability to print, and those from proprietary vendors of 3D models. These all should include some indication of the creation date of the model, but it may not include versioning control or updates to the model page for new or older versions.

Relevance: Google is intended for all audiences, and results must be reviewed individually for intended audiences.

Authority and Accuracy: As librarians know, Google searches bring back all sorts of results. Determining the authority and accuracy is based on the individual result.

Model Details: All model types are possible when using Google. There is no limiting factor that would exclude any model type.

Search Features: Keywords are the primary methods for searching, with some advanced features available. Search strategies for Google are listed next for more help in using this resource.

Be aware that some Google results may be blocked by IT due to the websites they are on; sometimes the model websites are flagged by IT security settings, and other times the results may actually be harmful, so being blocked is for the best. Google results incorporate artistic interpretations and, as a result, will add these to the number of results you need to review. Searching Google is time-consuming at first, but once you learn about reputable sources, it will be easier and faster. The following are some search strategies that may help during initial Google searches for 3D models.

Google Search Strategies

As a librarian, it can be extraordinarily difficult to tell someone you cannot find any results. This can happen in normal information searches, but when searching for 3D models, it will most likely happen more often. With that being said, more resources for 3D models are being created and shared online. The strategies here will help to narrow and focus searches completed in Google and to filter out irrelevant results.

File Extensions: When searching for 3D models online, one frustrating thing you will encounter is finding virtual 3D models in your results or even images and news stories about the 3D-printed models. Similarly, sometimes your results will include professionally manufactured 3D models from online retailers that are already made and not something that can be printed. One strategy to avoid these issues is to include file extensions in your search query. The two most popular file types for 3D models are STL and OBJ. Adding these to your search will filter results and yield a higher relevancy.

Multiple Keywords: As you search for models, test out adding a variety of terms for not only the subject of the model but also the end product to help steer the results toward printable models. Variations of the same subject should also be tested out, such as *3D* and *3-D*.

Truncation: One tricky aspect when searching for 3D models is making sure you include every variation of a term. Searching for 3D-printable models will be limiting with the word *printable*. Doing a more advanced search and using truncation will bring in results with variations in the terminology being used. Shortening phrases to the bare minimum will pull in all varieties of the phrase.

Advanced Search: Don't forget about Google's "Advanced Search" feature. Here you can attempt to refine your search with more options for combining terms rather than the single search element option. One note: The advanced search does offer an option to search by file type; however, 3D model file types are not included in the list and should be keyworded into the search box.

Test and Repeat: Just as sensitivity and specificity can alter results in a clinical query, using all of these strategies at once can be potentially too limiting. If your results are not what you expected, try dropping elements to see how your results are affected. There is a lot of trial and error when searching Google for 3D models, and it requires some patience when getting started.

NIH 3D Print Exchange (https://3dprint.nih.gov)

Background

The NIH 3D Print Exchange launched in June 2014 as a result of collaboration between the National Institutes of Health (NIH) and the National Institute of Allergy and Infectious Diseases (NIAID). The Eunice Kennedy Shriver National Institute of Child Health and Human Development (NICHHD) and the National Library of Medicine also help to keep the exchange online. A goal of the exchange is to "provide an outlet for creating and sharing medical models to facilitate visualization and learning" (US Department of Health and Human Services, National Institutes of Health, n.d.).

What's Included

Models include medical tools and devices, anatomy, lab ware, molecules, proteins, bacteria, and cells. Additionally, there are four special collections that highlight prosthetics, neuroscience, a heart library, and a molecule of the month. Typically, model pages include attribution information, instructions for how to print, model details, and general information. Instructions on how to print may sometimes be revised, and users can comment on the model page about any changes or updates needed.

The exchange also provides workflows and software options for users to create their own models. Users can create small molecules, biomacromolecules, image stacks, and models from medical scans. Tutorials are coming soon to the website to enable more users to gain the skills necessary to create their own models.

Evaluation

Currency: Publication dates are included, and updated model versions are available, with new version dates.

Relevance: The exchange is available to the public, but the primary audience includes teachers, students, researchers, and those with a passion for 3D printing.

Authority and Accuracy: The exchange includes an "NIH Verified" seal to indicate when a model has been reviewed by the NIH for accuracy. Additionally, users can submit comments, reviews of models, and inaccuracies if found.

Purpose: In the exchange's terms and conditions, it very clearly states the site is intended for educational purposes and not meant for medical use.

Search Features: There is basic keyword searching, with the option to include a category. Selecting a category with your search is recommended for increased relevancy in your results.

Model Details: Models are more focused on pathology, molecular and biochemistry, and lab support.

Thingiverse (https://www.thingiverse.com)

Background

Thingiverse is a creation of 3D printer manufacturer MakerBot and was developed in 2008 to help build a 3D printing community of creators and printers.

What's Included

Models included in Thingiverse cover a broad range of items, from fashion items to toys and games. In the mix of all these models, anatomy and other health and medically related models can be found, but it requires some searching and review of results for relevant models. In addition to models, educational materials are also included and provide lesson plans and activities for a variety of subjects (mainly K–12). Apps are also available and allow for model customization and outsourcing for 3D printing.

Evaluation

Currency: Models in Thingiverse have the option to allow other users to edit certain models. Thingiverse includes remixes of models and allows creators to include updates to the model pages. Fellow creators are often very good at using the "Comments" section to note if a newer or older version of a model exists.

Relevance: The intended audience for Thingiverse is broad, and as such, models must be reviewed individually for relevancy. Typically, creators include with models background information that will help determine the audience.

Authority and Accuracy: Thingiverse models are subject to the information provided by the creator to determine their authority and accuracy. Creators may simply name a model and not include any other relevant information, and as such, it requires consulting other sources to determine accuracy. In these events, sharing links to requestors includes them in the vetting process and encourages them to use their authority to select a model.

Search Features: Thingiverse offers a basic keyword search option and the ability to filter by category. There is no health or medicine category, but the biology option works well for anatomy.

Model Details: All sorts of models can be found in Thingiverse: normal anatomy, pathology, molecular and biochemistry, and lab ware. Furthermore, Thingiverse also provides stress-reliever models that provide an opportunity for students to test out the capabilities of 3D printing.

THINGS TO CONSIDER

Keep track of the models you find and the resources that prove to be useful. Creating your own list of resources to consult speeds up the request process. Chapter 8 highlights various methods to use to keep track of your 3D printing data and incorporate model files.

With 3D printing a growing trend in medicine, it is safe to say that more resources will be available. In chapter 14, social media accounts are listed and provide a good outlet for staying up to date on the latest information on 3D printing.

SUMMARY

- Remember to review resources for 3D models using the CRAAP method.
- Know what type of model you are looking for to identify a ready-reference source.
- Model types include normal anatomy, pathology, molecular and biochemistry, lab support, and stress relievers.

REFERENCES

Blakeslee, S. 2004. "The CRAAP Test." *LOEX Quarterly* 31, no. 3: 6–7. http://commons.emich.edu/cgi/viewcontent.cgi?article=1009&context=loexquarterly.

BodyParts3D. n.d. "Information." Accessed June 1, 2018. http://lifesciencedb.jp/bp3d/info/index.html.

Embodi3D. n.d. "About Us." Embodi3D.com. Accessed June 1, 2018. https://www.embodi3d.com/about-us.

Hanna, M. 2016. "An Interview with Dr. Michael Itagaki, Founder of the First Online 3D Printing Biomedical Community." 3D Printing Industry. May 9, 2016. https://3dprintingindustry.com/news/interview-dr-michael-itagaki-founder-first-online-3d-printing-biomedical-community-78490.

Mitsuhashi, N., K. Fujieda, T. Tamura, S. Kawamoto, T. Takagi, and K. Okubo. 2009. "BodyParts3D: 3D Structure Database for Anatomical Concepts." *Nucleic Acids Research* 37, database issue: D782–85. DOI: 10.1093/nar/gkn613.

Thingiverse. 2018. "About." https://www.thingiverse.com/about.

US Department of Health and Human Services, National Institutes of Health. n.d. "3D Prints in Medicine." Accessed June 1, 2018. https://3dprint.nih.gov/about/medicine.

Chapter Seven

3D Printing from DICOM Data

Creating models from digital imaging and communications in medicine (DICOM) is becoming more accessible due to both open-source and proprietary software releases. Interestingly enough, DICOM was created as a means to make sharing medical images easier, and now it is going a step beyond and making physical representations more sharable (Bidgood, Morii, Prior, and Van Sychle, 1997). 3D models printed from DICOM can provide surgeons with a means to better plan surgeries and better explain those surgeries to their patients. 3D-printing models from DICOM can immediately benefit health care and also can enhance medical education by providing a new medium for pathology and anatomy.

WHAT IS DICOM?

DICOM can come from magnetic resonance imaging (MRI) and computed tomography (CT) scans, as well as from ultrasound imaging and X-ray angiography (National Electrical Manufacturers Association, 2018). Printing from DICOM can create patient-specific models for planning and carrying out treatment. It is accessible to both health-care professionals and patients and has already made great impacts in patient care. One unique use of DICOM printing came in 2015, when a blind woman's doctor arranged for a 3D print of the woman's unborn baby created from the DICOM from the ultrasound (CBS News, 2015). Other uses of DICOM for patient care include generating models in order to make custom prostheses.

THE PROCESS

Printing a model using DICOM data for the design has benefits and drawbacks. Some of the benefits include genuine, realistic models that mirror actual human anatomy. This can be useful for creating study aids and providing real-life anatomy replicas on demand. Additionally, anatomical defects and damages can be printed to display the impact of diseases and help provide new vantage points in diagnosing conditions and aid in learning pathology. One drawback of using DICOM data is the need for the cleanup to make the model printable. This varies from model to model, but DICOM data that is not high resolution results in holes in the model that need to be repaired or removed for it to print successfully. Proprietary software enables easy and automatic model cleanup, but this is not generally included with open-source programs. Aside from the model cleanup, the complexities and technical skills needed to operate the software make using it difficult. As software improves, this process will become easier; currently, there are numerous resources online to support use of these programs, but time is still needed for learning, and the learning curve depends on users' technology skills.

TECHNICAL DETAILS

When working with DICOM data and creating a 3D-printable model, there are a lot of new terms and processes to learn. Proprietary software might spare you from the technical jargon or have such a simplified process that you do not need to understand the technical terms. Learning the terminology associated with DICOM data is one reason potential users might shy away from it when deciding to use your 3D printing service.

The following are some technical details when dealing with DICOM data:

Basic Image-Editing Tools: contrast, brightness, color keys, transparency

Data Probe: data about the position of the mouse in the scene

Decimate: reduces the number of polygons and triangles

Modules: functions set for manipulating the data

Nearly Raw Raster Data (NRRD) File: file formatting for image-processing applications

Noise: data in a model that distracts from the main model

Scene: location of loaded data

Smooth: to even out the model's surface texture

Structure: parts of the anatomy for modeling (e.g., brain, skull)

Threshold: upper and lower values that allow for certain densities of materials to be seen

This is not an extensive list of terms, but you most likely will encounter them when using DICOM-editing software. Luckily, there are a lot of tutorials and instructions available for DICOM-editing software, and knowledge of these terms is not completely necessary; however, when adjusting them in the software, it is good to know what exactly is being altered (3DSlicer, 2018).

DICOM SOFTWARE

When reviewing software to use with 3D printing, apply the CRAAP criteria to the program you are reviewing. Open-source software waxes and wanes and can have a lot of momentum when it first starts, but it may stop receiving updates and, as a result, be clunky to use. Reviewing the latest updates and additional information about creators will help you decide if the software will, first, do the job and, second, function. Nothing stings worse than finding the perfect software to solve a problem and then discovering that it is not compatible with your current operating system or is riddled with bugs. The following overviews of software, both open-source and proprietary, highlight their features and provide information on the ease of use and functionality.

Open-Source Programs

There are some options available for open-source programs that view DICOM data and export files for 3D printing. While the abundance of open-source programs enables this service for users, updates of these programs rely solely on the creators. Programs can be buggy and cumbersome, depending again on the creators. With the increasing need for this service, more steady and reliable programs have been developed and are listed here.

3DSlicer

3DSlicer is an open-source software that allows for DICOM data sets to be imported and manipulated to export an STL or OBJ file. DICOM data can be converted into a 3D model and then sliced and segmented to print a specific part of the anatomy, or the entire anatomy can be converted to a 3D model for printing or further editing in another program.

Currency: The most recent version, 4.8.1, was released in October 2017.

Relevancy: The intended audience is medical professionals and researchers.

Authority and Accuracy: 3DSlicer was created through grants made available through the National Institutes of Health (NIH). It was the result of a collaboration between various researchers from multiple institutions. The creators have since published on the development of 3DSlicer, which is available on PubMed (Fedorov et al., 2012).

Purpose: 3DSlicer is explicit in what it should be used for, and as soon as it loads, a pop-up window states on the "Introduction" page, "Slicer is not approved for clinical use and intended for research" (3DSlicer, 2017).

Ease of Use: 3DSlicer is a program that requires some technical skills and knowledge. Tutorials and forums are available for support, but it still requires time to learn how to use it. Numerous YouTube videos are available that provide good run-throughs on editing DICOM and step-by-step guidance. Once steps are learned, it is more streamlined and easier to use. The more advanced features will be what is most difficult to learn and require more time and training.

Democratiz3D

Democratiz3D is a program available through Embodi3D. Embodi3D is one of the featured resources for finding 3D models and, as such, provides Democratiz3D with credibility. Knowing that it is also the program used to generate models found on Embodi3D is also promising, as the results can be seen clearly. Furthermore, Democratiz3D also requires the use of another program to convert CT DICOM files to NRRD files. MRI files are not compatible with Democratiz3D. However, Democratiz3D offers a faster, simpler approach to creating 3D models from DICOM and can be completed in three steps: convert to an NRRD file, upload the NRRD file and select operation and license type, and fine-tune the model. More advanced users can set their own threshold limits, but this is not required, and you can use presets for bone, skin, or muscle. 3DSlicer allows for more fine-tuning in segmenting models, but Democratiz3D is a lower-barrier tool; it is easier than 3DSlicer and allows for bone, skin, muscle, and vascular models to be generated. Beyond these model options, users will need to explore more advanced software.

Currency: Being cloud-based, there are no versions to consider, and the most up-to-date version is always available.

Relevancy: The intended audience is a mixture of medical and health professionals and potential others interested in generating their own 3D models from medical scans. In the "Getting Started" section, the program is stated as being available to "Anyone who wants to quickly and easily convert medical scan images into 3D print ready files" (Embodi3D, 2018).

Authority and Accuracy: This resource was developed in part by a practicing medical doctor, which provides credibility to the software.

Purpose: The purpose of Democratiz3D is to enable easier and faster creation of 3D-printable models from DICOM data for sharing and collaboration. While the program is listed as being HIPAA-compliant, it does not address use in the clinical setting, and it is up to the user what they do with their model.

ITK-SNAP

ITK-SNAP is a medical-imaging software for producing segmented images. While 3D printing is not the focus, it does offer this option with exporting segmented models.

Currency: The most recent release was in April 2017.

Relevancy: The home page of the website includes announcements, one of which is a training session held for the Radiology Society of North America. This reinforces that their audience is indeed medical professionals.

Authority and Accuracy: Dr. Paul Yushevich from the Penn Image Computing and Science Laboratory at the University of Pennsylvania and Dr. Guido Gerig from the Scientific Computing and Imaging Institute at the University of Utah created ITK-SNAP. The focus in imaging from both creators provides authority to the software.

Purpose: ITK-SNAP is used for segmenting medical imaging. 3D printing is not the primary intended use, but again, as with other programs, this is possible, and models can be exported.

Medical Imaging Interaction Toolkit (MITK)

MITK is a software for enhanced medical imaging and visualizations. It is primarily geared toward virtual medical imaging for collaborative purposes. It does allow users to save models as STLs which enables it to be used for 3D printing.

Currency: The newest release of this software was July 2017.

Relevancy: The intended audience is broad, and this program is aimed at anyone with a need for creating interactive imaging from medical-imaging sources.

Authority and Accuracy: Created by the German Cancer Research Center, MITK–Diffusion projects and publications are highlighted based on use of this program. A PubMed search also sheds light onto the authority and accuracy, as it is a known resource in the medical-imaging field and is found to be very useful and beneficial.

Purpose: The sole purpose of MITK is to enhance medical imaging. Once launched, the software states that is it not to be used for diagnosis and treatment of patients. Creating 3D-printable files is not explicitly stated as an option with this software, but it is possible.

Proprietary Programs

As with the number of open-source programs, proprietary programs also have grown. The benefit of proprietary programs is the support services offered, and the functionality is usually easier to use. Prices vary, but free trials or demonstration versions are usually available and provide the option to test the ease of use and functionality of the software. These should be considered based on service and user needs.

DICOM to Print (D2P)

D2P is a software available through 3D Systems, which is FDA approved for creating 3D models. It provides automatic segmentation and streamlines the process of creating patient-specific models. Data from CTs, MRIs, and cone beam computed tomography (CBCT) can be used and set up to transfer print files directly to printers.

Currency: D2P is a recently released program from 3D Systems.

Relevancy: The product is marketed to the health-care community and includes expert testimonials.

Authority and Accuracy: The FDA approval provides credible support for the accuracy of the models produced.

Purpose: This is marketed for its patient-specific 3D-model rendering that is FDA approved. This implies that it is more for patient care but can still be used for educational purposes.

Conceptualiz—Ossa

Conceptualiz is a relatively new company, formed in 2012, that provides software to generate patient-specific, 3D-printable files. Marketing itself as an in-house medical 3D-printing company, it promotes the use of printing models, guides, and implants.

Currency: Being a proprietary software, currency is expected, but it is not clear if new versions or updates have been released.

Relevancy: The primary audience is health-care professionals, with surgeons mentioned specially for surgical planning.

Authority and Accuracy: The product does not appear to have FDA approval. It does have a clinical advisory board that provides some background on the authority of the models.

Purpose: It is clear that this product is for generating 3D-printing models. On the support page, use is listed as for educational and research purposes only.

OsiriX

OsiriX is touted as the most commonly used DICOM viewer worldwide. It is FDA approved and listed as a class II device for diagnostic imaging (OsiriX, 2018).

Currency: Being such a well-known product, currency is not an issue, as there is a large user market to keep the software current.

Relevancy: The product is geared toward health-care professionals. Feedback from doctors is also included on the product's page.

Authority and Accuracy: FDA clearance provides the authority and accuracy for the image quality and the models that can be exported.

Purpose: The primary purpose of OsiriX is as a DICOM viewer. 3D printing is more of a secondary feature and is made possible by exporting STLs.

Other programs, both proprietary and open-source, are available for generating 3D models, but research did not provide enough information to demonstrate that the resources are authoritative. Other programs might not have provided enough information to determine if 3D-printable models were able to be exported.

Printing from DICOM can be easy depending on what is desired. Bone and muscle can be done using open-source software but may still require some training. More advanced segmentations can be done but require additional training that might be a good investment depending on a user's needs.

REFERENCES

3DSlicer. 2017. "Documentation/4.8/Announcements." October 27, 2017. https://www.slicer.org/wiki/Documentation/4.8/Announcements.

———. 2018. "Documentation/Nightly/Training." June 22, 2018. https://www.slicer.org/wiki/Documentation/Nightly/Training.

3D Systems. 2018. "D2P (DICOM to Print)." https://www.3dsystems.com/dicom-to-print.

Bidgood, W. D., S. C. Horii, F. W. Prior, and D. E. Van Syckle. 1997. "Understanding and Using DICOM, the Data Interchange Standard for Biomedical Imaging." *Journal of the American Medical Informatics Association* 4, no. 3: 199–212.

CBS News. 2015. "3D-Printed Ultrasound Lets Blind Mom 'See' Unborn Baby." May 7, 2015. https://www.cbsnews.com/news/3d-printed-ultrasound-lets-blind-mom-see-unborn-baby.

Embodi3D. 2018. "Democratiz3D Frequently Asked Questions." June 1, 2018. https://www.embodi3d.com/democratiz3D-faqs/#whodem.

Fedorov, A., R. Beichel, J. Kalpathy-Cramer, J. Finet, J. C. Fillion-Robin, S. Pujol, C. Bauer, D. Jennings, F. Fennessy, M. Sonka, J. Buatti, S. Aylward, J. V. Miller, S. Pieper, and R. Kikinis. 2012. "3D Slicer as an Image Computing Platform for the Quantitative Imaging Network." *Magnetic Resonance Imaging* 30, no. 9: 1323–41. DOI: 10.1016/j.mri.2012.05.001.

ITK-SNAP. 2018. "ITK-SNAP." April 27, 2018. http://www.itksnap.org/pmwiki/pmwiki.php.

MITK. 2018. "The Medical Imaging Interaction Toolkit (MITK)." January 22, 2018. http://mitk.org/wiki/MITK.

National Electrical Manufacturers Association. 2018. "History." https://www.dicomstandard.org/history.

OsiriX. 2018. "OsiriX Features." June 1, 2018. https://www.osirix-viewer.com/resources/technical-sheet.

Chapter Eight

Data Management

When I first started 3D printing, I had no idea the amount of data that was generated until one day when I sat down with our newly hired data services librarian, Erin Foster. Erin and I discussed the library's 3D printing service in more detail, and it was at this time that I realized how much data we had and were losing. I wished I had a process or system in place for collection and storage. Attempting to establish this after the fact was difficult and is an ongoing process. To help other librarians interested in 3D printing, I decided to share what I learned and provide some suggestions for preparing and handling this aspect of 3D printing.

Whether printing anatomy or research-based models, 3D printing generates a lot of data that is not always immediately realized. Creating a print file is one piece of data that contains more data inside. The model itself contains data for size, orientation, and other adjustments that might have been made to the original design. Once the model is turned into a print file, even more data is created. If a print fails and needs to be adjusted or just needs to be replicated, it is important to have access to this data to correct a problem or simply reprint a successful model. Accidently reprinting a model with settings that previously failed is not only a waste of material but a waste of time, as well. It is important to develop a plan to manage the data that comes along with 3D printing, as it will save you time in the long run when problem solving and for repeat print requests.

WHAT IS DATA?

Ray (2014) identifies two ways libraries can help with data: management and curation. While data curation focuses more on increasing data's availability to external users, data management focuses on internal accessibility. Data management is a popular, fast-growing area of librarianship. For librarians new to this subject, data management can be overwhelming if the basics are not well understood. To understand data management, you must understand what exactly constitutes data. When starting my library science degree, my very first class had a similar exercise, in which information was discussed and how we defined it. *Information* can best be defined as that which gives us knowledge. When considering data, it, too, can fall under this definition. Information and data both provide users with knowledge; however, data diverges from information in that it needs information to be understood. *Data*, as defined by the School of Data, is the "value of a thing" (2013). On its own, data may just be a series of random numbers or words, but adding background information to data allows for conclusions to be made regarding the answers the data is providing.

While it has one definition, *data* can have different meanings to different groups and organizations. For instance, in business, data might be in relation to customer shopping habits and marketing trends. Data can be pulled from loyalty card transactions or discovered through surveys and interviews. In academia, data might be more research focused and not provide insights into trends as much as it provides answers to hypotheses and theories. Data can be generated from an experiment and, as such, requires a level of strictness in how it is gathered and stored. Furthermore, these different uses of data fall into separate categories: qualitative (the quality of an object); quantitative (the amount of objects available); categorical (the category of the object); continuous (numeric values associated with an object that can be measured on a scale or continuum); and discrete (an object that has set values and gaps between those values; School of Data, 2013).

Gathering and storing data is something discussed in a data management plan. 3D printing falls somewhere between academia and business in regard to its data. Not only can the data provide information on trends and users, but it can also provide more internal information on the service itself. Creating a data management plan prior to establishing a 3D printing service allows for a process of data collection that provides a good overview and understanding of how the service is operating. Even if the service is already established, instituting a data management plan still provides good insights and can help guide the service.

3D PRINTING DATA

Prior to discussing a data management plan, it is important to first understand the data produced through 3D printing. When you first begin 3D printing, you are not aware of just how much data is generated. 3D printing creates a large amount of data, if you know where to look for it. The following terms are some areas where data is generated in 3D printing (*3D Printing for Beginners*, 2016):

Model Weight: how much filament is being used
Print Time: how long the print will take to complete (experience has taught me that this fluctuates)
Infill: how models are filled in (solid, diamond or honeycomb pattern, lines)
Triangles/Shells: how detailed the model being printed is (more curves means more triangles and shells)
Print Speed: how fast the extruder ejects filament
Resolution/Layer Height: smaller layer heights equal a higher resolution but a longer print time
Wall Thickness: the thickness of the outer layer of the model
Extruder Temperature: the temperature used to melt filament down
Filament Type: PLA, ABS, other
Support Density/Angle: type of support and amount of support added to model for printing

The information gained from 3D printing data varies but ultimately will provide insights that lead to better print operations. Understanding the data that is generated is one thing, but understanding how it affects final models is another. Fully understanding each element will better help you decide the data you choose to collect:

Model Weight: Lighter-weight models will save in material costs, but weight can also affect print time. Additionally, model weight may be divided between model weight and support weight. It is useful to know if material is being used that will ultimately be discarded and not go toward the final product. This is important to measure, as it will help determine material budget and possibly cause you to try alternate orientations to reduce the support weight. Finally, unless your printer uses a sensor to determine how much material is remaining, keeping track of the model weight enables you to determine when filament will run out and need to be replaced.

Print Time: Print times can help determine turn around time and organize a print queue. It may also provide background information on printer maintenance and identify when servicing will need to be completed.

Infill: Infill can have a big impact on print time and model weight. Infill is generally described as the strength of the model, and keeping track of the type of infill and the results on the final model (e.g., Does it break coming off the plate? Does it withstand support being removed?) is helpful.

Shells: Shells are the outer-facing walls of 3D models. If shells are too thin, the model can develop holes in printing. Furthermore, if they are too thick, it could hamper alternate uses, such as translucency.

Print Speed: Adjusting the print speed may result in a faster print but may also result in a failed print or a print with such problems as quality issues.

Resolution/Layer Height: Layer height affects print time, as smaller layers require more material and time. Higher resolutions are better for more detail and may be necessary for some prints, depending on the final purpose of the model.

Wall Thickness: The thinness of a wall may result in a fragile model or one that cannot withstand support being removed. Similarly, depending on the filament type (e.g., translucent), a thinner wall is required if transparency is the goal. Printing in a flexible material also requires special attention to the wall thickness, and monitoring successes and failures enables better print outcomes in the future.

Extruder Temperature: Default temperatures may yield no issues during printing; however, experimenting with different temperatures may provide good insights on alternate temperature settings that will produce better print outcomes.

Filament Type: Filament type can have a variety of impacts on the final model. ABS may result in fewer failed prints, or PLA may result in a higher quality. Comparing model types and outcomes with the different filaments helps determine (if it's an option) the type of filament to use with print requests.

Support Density/Angle: Adjusting the support density and angles helps provide intricate models with better outcomes, and adjustments might also help to reduce model weight if this is an issue.

USES OF 3D PRINTING DATA

Whether you decide to collect it is entirely up to you. Inferences can be drawn from the data but might not be readily seen until it is analyzed. For instance, if you do a little more experimenting in your service, over time, you might notice that the filament type shows trends with better printing at certain extruder temperatures. Furthermore, you might also find a combination for print speed and wall thickness that increases the resolution without adding to the overall time. You can also use data to predict potential maintenance for the printer. For example, extruders sometimes have a "shelf life" or a guarantee to print problem free until a certain number of hours is reached. Keeping track of the hours each extruder has printed can indicate if one extruder might experience problems and require replacement or maintenance.

Karel Brans of Materialise discusses the need for data management in 3D printing. Geared toward manufacturers, the need described by Brans still applies to other industries. Manufacturers who use 3D printing need to consider storing the data associated with digital models for quality assurance. Brans also highlights medical manufacturing and its own data needs for devices. Medical devices, in particular, require data to be stored for up to seventy years (Brans, 2013). Promoting good data management practices early in printing will help guide new users into following similar practices.

Another data need Bran highlights is "traceability." Traceability is tracking which printer prints a certain model. One way manufacturers do this is to engrave the digital model file with the printer number. If a model fails, then the printer can be identified quickly using this number. This should not be as issue for libraries printing on a much smaller scale; however, it might be a useful feature to track the durability of models over time.

While librarians traditionally provide researchers with information and guidance on data management plans, in the case of 3D printing in the library, it is a case of practicing what you preach. DAMA International, a community of data and information professionals, defines *data management* as the "development, execution and supervision of plans, policies, programs and practices that control, protect, deliver and enhance the value of data and information assets" (Dennis, 2017). Capturing the data associated with 3D printing will provide libraries the ability to not only track usage but also determine how their machines are operating, predict filament expenditure, and anticipate printer maintenance.

DATA MANAGEMENT PLANS

Knowing how much data is produced can be overwhelming when thinking of how to track everything without getting bogged down in record keeping. If you are starting a data management plan after already starting a print lab, it might be difficult to get in the process of collecting this data, especially during busy times. However, you also have the advantage in that you've been printing already and will have a better understanding of what might be most important to collect based on previous requests. A recent editorial published in *Nature* emphasized the importance of creating a thorough data management plan to draw meaningful results and to be able to reproduce results (*Nature*, 2018). This is especially important in 3D-printing temperamental models that may have already undergone a series of trials and errors and require very specific settings for success.

Data management plans do not need to be extensive; they should cover four main elements: defining what your data is; identifying potential ethical and intellectual property rights issues; determining how you will store and access this information; and identifying the tools and resources available to help in your plan (Krier and Strasser, 2014). Earlier in this chapter, the data needed is defined, and in chapter 2, legal and ethical issues are discussed. Tools and resources are also discussed in the next section. There are options available through open-source programs to help build a data management plan. These programs are more tailored to research-focused projects but can guide you through the steps in developing a data management plan for a 3D printing service. DMPTool is one popular resource for researchers and can be used for establishing a 3D printing service data management plan. DMPTool and additional programs are featured in chapter 14.

DATA MANAGEMENT SYSTEMS

Managing and storing the data associated with 3D printing is dependent on the service. If you are crunched for and lack staff, you might be able to collect less data. To get started identifying tools to assist in managing your data, proprietary software is reviewed. The services' capabilities and costs are also discussed. These are programs that are very dependent on your 3D printing service and the needs of your users. Knowing about these services ahead of time allows you to include the cost in your budget if you decide that any of these options are useful.

3DPrinterOS

This system seems like the best option for a library print lab as it offers a case study, "3D Printer Management for Education," on its home page. 3DPrinterOS offers features for schools and universities and includes software that enables editing, repairing, and slicing to create print-ready files and the ability to "audit" any step of the printing process. Additionally, having an all-in-one software allows for the simplified preparation of print files and can eliminate the need to export between programs.

There are four plans available, ranging from a free starter plan to a $200-per-user premium plan. Two other plan options, educational and enterprise, only list "Contact Us" as an option for pricing. The starter plan offers the ability to upload and print GCodes and monitor remotely. Forum support is also available. The premium plan includes the option to queue prints and share printers, projects, and files. Live support for help is also available with this plan. The educational plan offers the machine data reports that are useful for analyzing the service in more detail. The enterprise plan seems to be aimed toward businesses and professional services and may have more features available, such as customized workflows and a direct print API, which allows you to us a web-based platform and print from anywhere. At this time, all plans include a version control for projects and files, which makes the starter plan worth investigating.

FabPilot

FabPilot is similar to 3DPrinterOS in that it provides model editing and print-file generation in one program and includes workflow management and quality reports. FabPilot is a product of Sculpteo, a 3D printing company.

Pricing ranges from $99 to $199 per month. Unfortunately, to learn more about the details of each plan, a consultation is needed. Another major drawback at this time is that FDM printers are not yet integrated in the software. It is in the works and is one reason I include this system. A fifteen-day free trial is also available for libraries wishing to test it out.

Cura Connect

Cura Connect offers users the ability to control their printers, track print status, and start prints automatically. Additionally, multiple printers can be grouped together and the print queue shared between them. While Cura Connect is available at no cost, it is only compatible with the Ultimaker 3 printer. Ultimaker printers are growing in popularity, and this might be an option for libraries that are considering this printer or already have these printers.

Materialise Streamics

Materialise Streamics is geared more toward manufacturing, as the product is marketed to "manage and streamline your AM activities" (Materialise, 2018). Streamics monitors the day-to-day processes involved with 3D printing and can determine printer availability based on model size and tracking used print-bed space. Data associated with printing is also stored within Streamics, and reports can be produced for further analysis. The cost of Streamics is not advertised, but trials may be available to see if this product might benefit your print lab.

The perk of using 3D-printing-specific software to manage the data is it is most likely automatic and streamlined, which, if you do have a busy lab, might be worth the cost. Manufacturing and professional services are those that will most benefit from this software. There are a few open-source resources available, as well as programs in Microsoft Office that can also help manage 3D printing data.

Microsoft Excel and Access are both easy-to-use options to create a system to manage your data. Whether you are an Excel fan or prefer Access, both allow you to create a data table that can then be used to analyze stored data. Furthermore, depending on your skills in either program, you can set conditional formatting to serve as alerts for filament usage. In Appendix B there is a list of fields to use in either Microsoft Excel or Access for data collection.

For more tech-savvy users, Git-Hub is gaining popularity in managing services like 3D printing. Another open-source tool, Git-Hub provides workflow management, versioning control, and storing files. Git-Hub also allows ReadMe files to be stored alongside print files and can serve as instructions for printing. Another benefit of using Git-Hub is the ability to preview the 3D models with a model-viewer tool. The one drawback to Git-Hub is that it can have a bit of a learning curve to get started.

REFERENCES

3D Printing for Beginners. 2016. "3D Printing Glossary." http://3dprintingforbeginners.com/glossary.

Brans, K. 2013. "3D Printing, a Maturing Technology." *IFAC Proceedings Volumes* 46, no. 7: 468–72. https://doi.org/10.3182/20130522-3-BR-4036.00112.

Dennis, A. L. 2017. "Data Management vs. Data Strategy: A Framework for Business Success." Dataversity. September 19, 2017. http://www.dataversity.net/data-management-vs-data-strategy-a-framework-for-business-success.

Krier, L., and C. A. Strasser. 2014. *Data Management for Libraries: A LITA Guide.* Chicago: ALA TechSource.

Materialise. 2018. "Materialise Streamics." https://www.materialise.com/en/software/streamics.

Nature. 2018. "Everyone Needs a Data-Management Plan." *Nature* 555, no. 7697: 286. DOI: 10.1038/d41586-018-03065-z.

Ray, J. M. 2014. *Research Data Management: Practical Strategies for Information Professionals*. West Lafayette, IN: Purdue University Press.

School of Data. 2013. "What Is Data?" September 2, 2013. https://schoolofdata.org/handbook/courses/what-is-data.

Chapter Nine

Getting Involved

Zen and the Art of 3D Printing

Once it is decided to start a 3D printing service, the real fun begins. Planning ahead of time will give you the best advantage to a smooth and successful service offering. Some of the areas to focus on when planning your service, or even revamping your service, revolve around the basics: how much to charge, policies, and the workflow for fulfilling requests.

3D PRINTING SERVICE: WEBSITE REVIEWS

There are multiple factors to consider with 3D printing services, such as where your library gets the funding for the service, who can use the service, and the type of printers you will use. One thing to consider when determining the pricing of the service is how you plan to operate the service. Will student employees manage and fill requests? Will postprocessing be included? How will models be delivered? Is mailing involved? Will you provide any sort of packaging? These are just some of the questions that will come up when planning your service. To get the best idea of what to do, it is important to see what others are doing. My earlier survey provided some insights; to go one step further, I reviewed hundreds of websites of academic libraries across the world with 3D printing services. To help with this process and ensure that I reviewed as many libraries as possible, Amanda Goodman's website, Map of 3D Printers in Libraries, provides a list of libraries offering 3D printing services (http://www.amandagoodman.com/3d).

In the end, I managed to review sixty websites of academic libraries in ten countries. Seeing how libraries offered this service across the world provided me with a new sense of awareness about the service. I reviewed their charging mechanisms, unique elements of their services, and the types of printers they owned. The following are the results from my website scan and the summaries of what I learned. A full listing of websites reviewed is located in the recommended resources list found in chapter 14.

Results

Countries: Australia, Canada, China, Denmark, Ireland, Japan, South Africa, Sweden, United Kingdom, and United States

Printers: FormLabs, LulzBot, MakerBot Replicator, MakerGear, RoboBeast, Stratasys, Ultimaker, Upbox, and ZYYX

Charging Mechanisms

Prices ranged from free to $0.04 to $0.20 per gram to $1.00 to $4.00 per hour. Also, some services include setup and processing fees ranging from $1.00 to $4.00. Prices varied per machine and material being used, and some libraries offered free or discounted academic-based prints.

Service Features

All services include copyright disclaimers and the library's right to refuse any print. Roughly 25 percent allowed students to use printers. All required mandatory training of some sort—either an hour-long workshop or reviewing a safety sheet and a quiz. Additionally, the time estimates ranged from two to three business days to at least one week to allow for printing. Furthermore, some print labs refused to let printers run overnight. Quality of the final print was also something addressed in many 3D printing policies, and some libraries included disclaimers in the request forms themselves to ensure users were aware of this issue. Libraries also noted that models that failed due to printer error would not be charged to requestors, but design failures would be charged. When it came to the requests themselves, some libraries made the printers available via appointment scheduling, and others offered them as a drop-in service. An interesting feature offered at one library was the ability to check out postprocessing tools. Furthermore, some libraries only allowed for PLA filament, citing potential health issues. Finally, there was a variety of additional features that library websites would offer to help users better understand their services. Some libraries included a status notification for the printers, and others included live webcams to see the processes. Libraries also promoted their prints and offered galleries for users to review.

Policies

On almost all the services' websites, the libraries clearly listed their policies for what is and is not allowed to be printed. Unfortunately, recent news has resulted in a new area of concern that libraries must also address. All the websites were explicit that weapons and illegal items could not be printed. This will be a growing concern due to the reversal of a previous government order that forbid the 3D files of a printable gun from being shared online.

The Liberator is most likely one of the fears administrators have about 3D printing—something being printed that could hurt someone. In 2013, the Liberator gun was first 3D-printed. Almost immediately, the US State Department forced the files to be taken offline. Since then, a recent court settlement has allowed them to be shared online on August 1, 2018 (Williams, 2018). This might be more concerning for print labs that are self-serve, but it is important for any 3D print lab to be aware of because, more than likely, you will be questioned about how you will prevent someone from printing this model. On the websites I reviewed, every lab explicitly states that no weapons can be printed using the service, but even with this statement, especially now with these files online at no cost, you will most likely need a secondary measure in place to ensure that this does not happen.

There is an easy way to learn more about files that you are sent and their intended use: Google. The first and usually only thing you might need to do is Google the file name. I recommend not including the file type (drop .stl or .obj). A lot of times, when I received requests with files included, the names were not changed from when they were downloaded. Granted, I never had to confirm I was not printing anything dangerous, but I was able to find where models were downloaded so I could review print instructions or read comments about potential problems.

If you have serious concerns about something you've been requested to print, the easy answer is to not print it. Another item to consider is additional requirements for users who submit files for printing. Adding a verification of where the models were found is something other academic library 3D print labs include. I initially did not put too much thought into this, thinking it might be a way to review print instructions. But as they are a broader service, it serves as a means to track the models' sources and confirm what they are used for. This is definitely something to consider for user-submitted files. If the files are original designs for prototypes, you could request more information on the background of their project. Hopefully this will never be an issue, but administrators will most likely have this concern and ask about it. Having a procedure in place for reviewing user-submitted files is better than scrambling to set one up after being prompted. Not only will it save you from having to think of one on the spot if the need suddenly arises, but it also will preempt any concerns from administrators and demonstrate that you are aware of the potential issues with 3D printing.

TIME FACTORS

3D printing is a time-consuming service. The amount of time required varies depending on how you choose to operate the service. There are a few different elements to the service that take time to complete and need to be considered when planning the service so that staffing is appropriate.

Print Interview

Now that there is a better understanding of options to consider for structuring the service, the next thing to review is the amount of time you should expect to devote to managing the service. Prior to printing anything, you should conduct a reference interview of sorts with users interested in printing. The print interview should get the specifics and intentions of the model from the user.

Questions to Ask

What do you want the model for? Is it for studying or a print just for fun?
What kind of material do you want your model printed in; flexible, rigid, or with such special features as transparency?
If you bring a model with you, where did you get the file? Did you design it yourself? If so, did you check it for errors? Is the scale set?
If you ask for a model to be found, do you have a size preference? (Most students want full size but seemed happy with smaller versions.)
How soon do they need the model? (A big issue I've found with 3D printing is the unawareness of the actual time it takes to print something.)

This process can be quick and more than likely completed through e-mail; however, users new to the technology might also want a little tour of the print lab, and the initial interaction you have is always a great opportunity to get to know the user and promote the service, as well as other aspects of the library.

Prepping the Print

Now that you've got your request, there are a few different ways your next step will go. You either will have to search for a model or go straight to prepping the model. Chapter 6 discusses finding models, and chapter 10 goes over editing models, which will help guide both processes. Furthermore, this process can take some additional time if the user is requesting a faster print time and you are trying to keep the model a reasonable size. Another factor is whether printers run overnight. If your service allows this, you might decide to stage the queue of prints so that they print in a logical and time-efficient manner. You also may consider loading the print plate if possible to complete numerous requests in one, long print. This can be a time-saving trick or a time-consuming waste, depending on if any printer errors occur, causing the whole plate to fail. This is also one major thing to consider when deciding whether to print overnight: What happens if a print fails? Some printers can detect when mechanical issues occur and stop the print. However, not all failures are mechanical, and as such, the printer will not detect these and continue until the print time ends. Some printers offer a built-in camera to view printing and include a remote start and stop feature to address these issues. Other alternatives of monitoring include setting up your own camera and finding a means to stop the printer remotely. Libraries open late might be able to have staff available to turn off the machines, while others might be able to set up a remote power box that can remotely activate a power source.

Printer Fails

What are some of the things that can go wrong when 3D printing? I will not lie: There is a lot that can go wrong. One of the best metaphors I learned from the 3D print group at Indiana University is that 3D printers are like toddlers. They can have breakdowns, throw tantrums, and not do what they are told. In a 3D printing workshop, I often compared 3D printers to puppies:

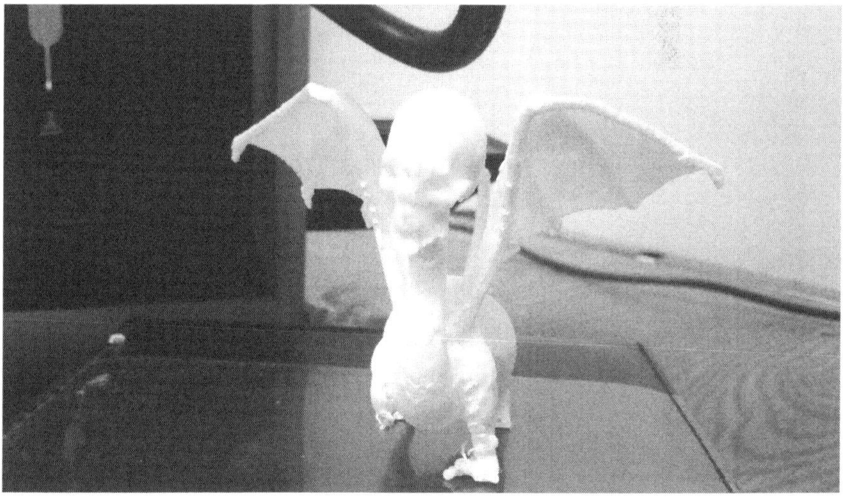

Figure 9.1. Fail dragon. Failing is in the eye of the beholder. ***Jennifer Herron.***

They are fun but destructive. Both toddlers and puppies get older, and as they do, they learn, and things settle down eventually; the same is true with 3D printers. You will begin to recognize problems before they happen and be able to deal with them before they destroy a model, and the hectic firsts with 3D printing will eventually settle down as you mature in your 3D printing skills.

The following are some of the common printer fails I encountered and the solutions I found to resolve them. For the solutions I provide, I did research prior to testing things out, and our printers did not seem to be affected negatively; however, if you have concerns, please consult with manufacturers for the best advice and solutions.

1. Clogged Extruder: The Most Encountered Problem

A clogged extruder can occur for a few reasons. If the Z-gap, or distance between the print bed and extruder nozzle, is too close, melted filament will jam back up into the nozzle and cool, at which time you develop a clog. Filament can also break off, and little sections can get jammed by the gears that push the filament through the extruder. This small section of filament can melt down and again, once cooled, cause a clogged extruder.

The frustrating thing about a clogged extruder is that it can occur at any time during the print. You might notice it right away when the printer does not extrude filament at the start of the printing process, or it might happen in the middle of print. I had an eight-hour print that failed in the last twenty-five minutes. What I have also learned is the more you fail, the more creative you will get, and the more you will learn about 3D printing.

To fix a clogged extruder, you can try to purge the material and see if this melts the clogged plastic, so it can work itself through the nozzle. There should be a purge setting on your printer for this problem. If purging does not work, some special equipment can be used. With the filament heated up (there should be manual option to raise the temperature on your machine), use a needle, guitar string, piano wire, or other similarly thin piece or wire to try to force the plastic through the nozzle. The printer manufacturer will most likely have a specific tool to recommend for the correct sizing to fit through the nozzle.

2. Model Does Not Stick to the Print Bed or Gets Ripped Off While Printing

This is a problem that will either be immediately known or you won't find out until you see the printer printing but with no model in sight. At the start of your print, if the filament sticks to the print bed but lifts as soon as the extruder changes direction, there might be a problem with the surface of the print bed. Another way this problem might make itself known is with taller printers that begin to lift from the print bed as the model gets taller and less

Figure 9.2. A clogged extruder can be especially frustrating if you have a dual extruder and one continues and the other quits. ***Jennifer Herron.***

structurally stable. An example of this happened with a knee model I printed. While it had supports, as soon as it reached the patella, it tipped over and fell completely off the print bed.

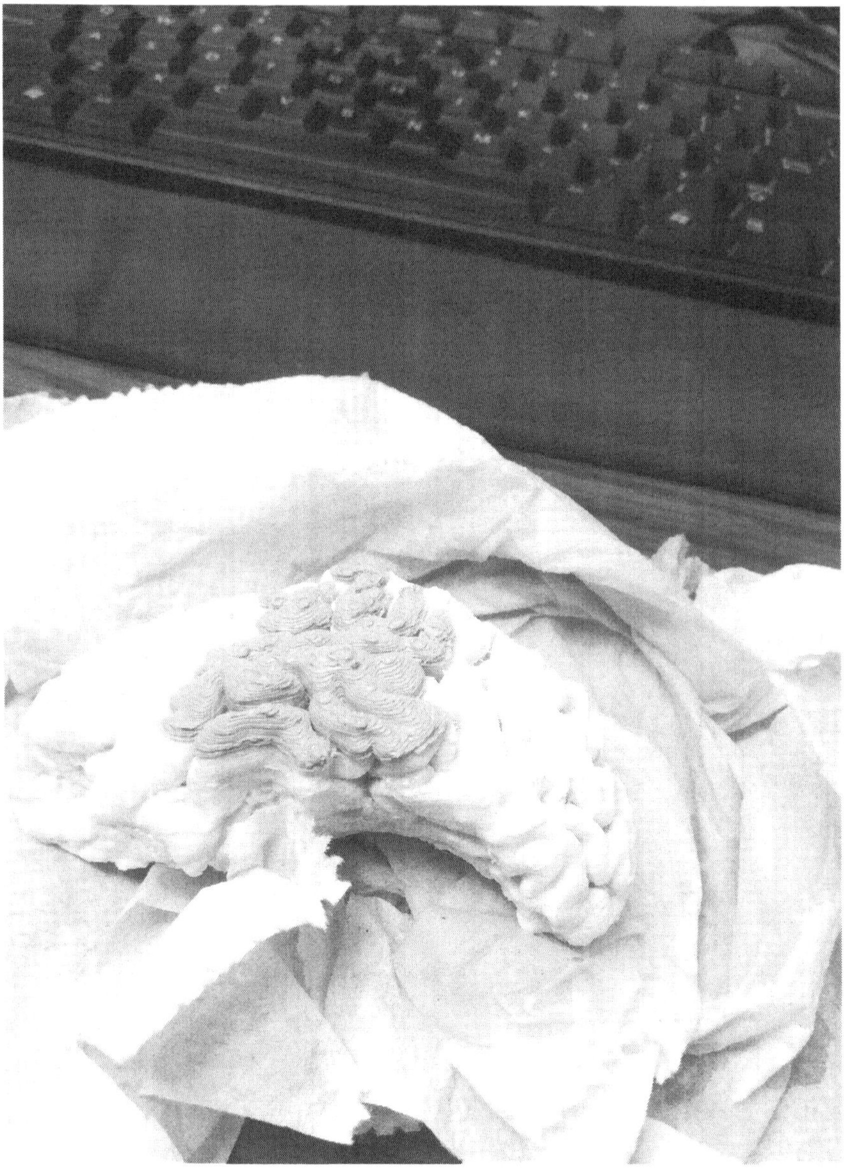

Figure 9.3. Sometimes the more you fail, the more creative you get. This brain stopped about twenty minutes before it finished—an eight-hour print total. This is where I learned to slice models. The final model received a new slice printed in neon green to match. *Jennifer Herron.*

The cause of these problems: There was not enough traction on the print bed to keep the model base rooted to the print bed. Some printers come with glue sticks for applying a layer of adhesive to the bed at the start of the print; if this step is skipped or if the glue has lost its stickiness, it will cause this problem. Blue tape or sheets of tape can be purchased to stick to the print bed, and these usually solve the problem. Another solution I found after doing some research is to spritz the print bed with hairspray.

Another culprit behind this issue could be the Z-gap. If the gap is too wide, the melted filament might cool before it hits the print bed and then not have the tackiness it needs to adhere to the bed. This issue is very dependent on the type of printer you have: If there is a heated print bed, this might not be an issue. However, with a wide Z-gap, the filament will still land too high and might not hit the bed correctly, causing it to shift.

3. Warping

Another issue you may encounter during the printing process is warping. Warping happens when the cooled plastic begins to curl away from the print bed. If you have an enclosed printer or a heated print bed, this might not be as much of an issue. While you cannot do much about not having a heated print bed, you can do something about the enclosure. I was able to repurpose the filament packaging and, using some electrical tape, cut the packaging to create a tented cover that trapped the heat and kept the plastic fixed to the bed. Printer enclosures can also be purchased, or instructions are available online from individuals who've built their own. Not only can enclosures help prevent warping, but they also can address any concerns about particulates emitted during printing.

4. Filament Breakage

One issue that is not preventable is when the filament breaks. Depending on the type of filament you purchase (encased or on a spool), you might have a better vantage point to see how the filament looks and if there are any creases or other signs of damage. Other times, a sign that the filament has broken is that material is not being extruded or the printer detects a problem and pauses the print. What's worse is if the filament breaks in two pieces and traps a small piece inside the extruder. This might require you to open the extruder head to retrieve the broken piece, but a word of caution: Some manufacturers will void the printer's warranty if you open the extruder head. If your warranty has run out and you need to open the extruder, I recommend checking YouTube for some how-to videos to walk you through the process. Another tip: Take pictures of the extruder as you take it apart. This will help you see how it was set up, so you can put it back together correctly.

The only real way to prevent filament breakage is to store your filament in airtight containers so that moisture and humidity don't affect it. Storage recommendations for filament include using vacuum bags with silica-gel packs and keeping it in a cool location away from sunlight (3D Printing for Beginners, 2018). Additional tips include using uncooked rice to absorb extra moisture in storage containers. Some creative solutions include altering airtight containers to allow for filament to feed in and out during the print process so that it can be exposed to as little air as possible.

When printing, a popping sound is an indication that your filament has been affected by moisture and may need to be replaced. Furthermore, one way to test PLA filament is to see how easily it breaks when bent. If it snaps, it is a sign that it may not be the best quality to use for printing (Ultimaker, 2018). If you suspect your filament has been compromised by moisture, there are some options to consider that may extend its life. One way to reduce moisture in filament is to dry it. Commercial filament dryers are available, and instructions are available online for using convection ovens (3D Printing for Beginners, 2018; Landry, 2016).

Another similar issue is when filament runs out. While this is definitely a much easier problem to fix, it does not make it any less frustrating if your printer does not detect a filament flow error and continues to print without extruding anything. If you are good with maintaining your statistics, you can make a filament tracker and input new print jobs to determine if you will have enough material to complete the print. If you know it will run out, you will have to stay close, so you can replace it and keep the print from pausing too long. While manufacturers claim their printers can remain paused for hours, I personally have had issues with alignment when the print started back up. In some cases, it was not too noticeable, and in other cases it ruined the print completely.

5. Mind the Gap

The Z-gap is mentioned previously as being the root cause of some problems; however, if it does not cause the extruder to clog or the material to hit the print bed too high, it might cause the extruder nozzle to drag itself through the material as it prints. Running calibration is recommended, and if there is an autocalibration, use a piece of paper to test the gap manually and make adjustments as necessary. When manually setting the Z-gap, you will gradually move the nozzle and print bed closer and set the paper between them. As the gap between the nozzle and bed gets narrower, the paper should just catch on the nozzle; at this point you can set the Z-gap. An important note: If you are printing with different material types, it is recommended that you recalibrate when you change filament. I often ran into problems when switching from ABS to PLA and found that recalibrating helped to solve many issues.

6. Update, Update, Update

Along with the printer software, the printers themselves need to be updated. Their firmware requires regular updates, and if not completed, your printer will start to show signs of needing an update with erratic prints and unusual fails that do not make sense. If your printer starts to fail and you cannot find an obvious cause, I recommend checking the firmware for updates. On this same note, I also recommend running through the calibration setup. The more your printer prints, the more things shift and move around, which might not cause too many problems with the prints immediately but over time will continue to get worse until you have a print fail.

One extremely useful piece of advice that I learned at a 3D printing workshop from a group of doctors who managed the 3D print lab at the Mayo Clinic is to print "coupons" that show you how your printer is functioning. I wish I knew this when starting our 3D printing service. At first, I searched manufacturers websites for such models but did not find anything. Thingi-

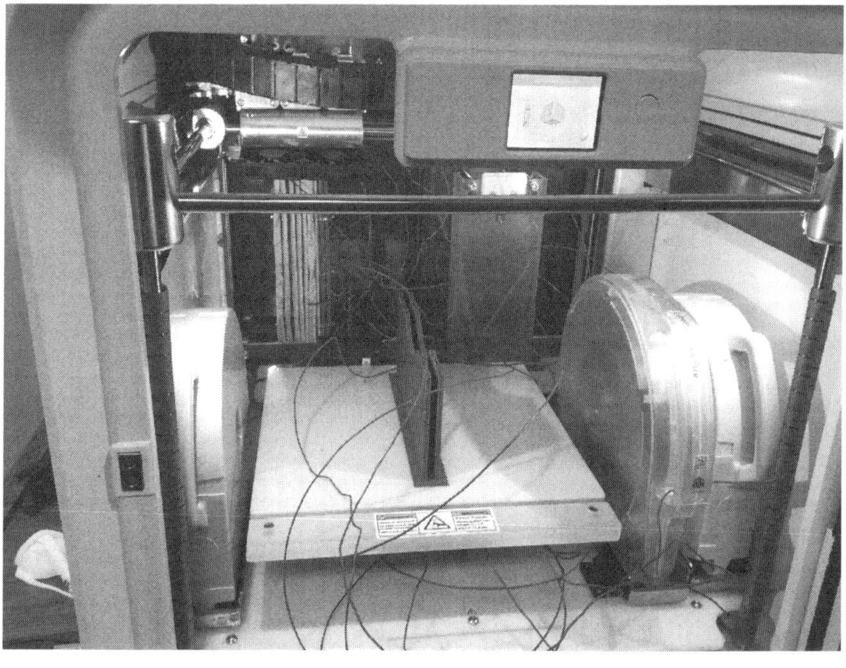

Figure 9.4. No clogged extruder, no filament break. No idea what went wrong. Calibrate, update, and try again. *Jennifer Herron.*

verse does provide such models and information on interpreting the print quality. Pleasc see chapter 14 for links to these models. These models can be large and take up almost the entire print bed, but there are also models that are designed to be smaller, so still test all the appropriate elements needed for calibration. One such model on Thingiverse is called the "Jolly 3D Printing Torture Test" by Creative Tools.se, which tests the printer while completing a model of a small boat. Other torture tests can be found online to put your printer through the wringer (https://www.thingiverse.com/thing:763622). Because URLs change and webpages can be deleted, these are search terms for Thingiverse in case any of the links break: calibration, printer test, and benchmarking. The group from the Mayo Clinic recommends doing this monthly to ensure your printer is operating correctly (Jennings, 2018).

Unfortunately, more printer problems than what I have listed can happen during 3D printing. It all depends on the printer you have and the model you are printing. YouTube and Google are excellent resources for troubleshooting printer problems. In chapter 14, I also highlight go-to websites for troubleshooting help.

DESIGN FAILS

Not all print fails can be blamed on the printer. Sometimes there are problems with the model design itself, and other times, it is operator error—a failure to set up the print file correctly before sending it to be printed. I do not have a background in 3D modeling and know there are likely other design errors that can be fixed in more appropriate ways than what I suggest here. As always, consult with experts in your area for best tips and advice. Problem solving is also a great networking opportunity! Don't have anyone in your area? Reach out online: Some resources for online 3D printing communities to contact for support are found in chapter 14.

Other design issues that can cause your prints to fail include wall thickness. Typically, when wall thickness causes a failed print, it is due to the walls being too thin. This can cause holes or gaps in the plastic. Adjusting the wall thickness can be done using the repair feature in some of the editing software. This can be a tricky fail to catch, as you might not realize it until after your model has printed and you can visibly see thinning and gaps in the plastic.

One strange issue I encountered a few times was when the extruders ran outside the normal print area and either to the sides or back of the printer, with a loud motor noise sounding until the printer was turned off. This error was a little more worrying, but it was easy to correct.

Before starting your 3D printing service, or even if you already a have a service, I recommend using another organization's 3D printing service. This

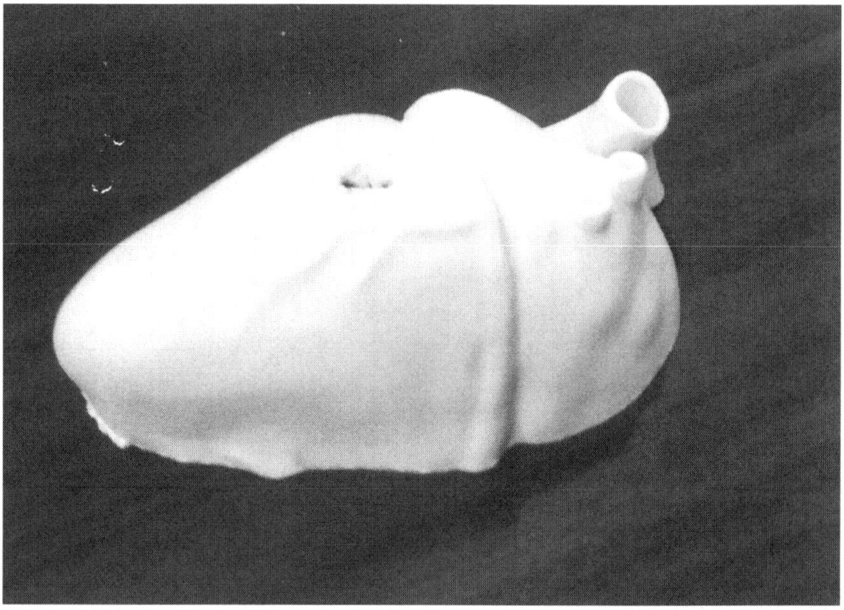

Figure 9.5. Operator failure. These sometimes do not show up until the very end. Forgetting supports can cause holes and other times a big mess before the model gets too far. *Jennifer Herron.*

could be another library or even a commercial 3D printing service. I recently did this and found it to be a great experience. Not only did I not have to worry about prepping my model or fixing any printer errors, but also I got to experience 3D printing from the other side. I was excited about my request (being back in Michigan, I chose the Mackinac Bridge). I was able go through the 3D printing process from the perspective of the user. Having managed the 3D print lab at the Ruth Lilly Medical Library, I had some ideas of what to expect. I also got the chance to see how other libraries managed their 3D print services. In this position, I was able to see what I thought went well in the interaction and what I would have done differently. Visiting an outside 3D printing service gives you the opportunity to see how you'd like to experience the process, and then you can use what you've learned as you develop your service.

ZEN AND 3D PRINTING

Whether you have student employees or other staff managing your printers or it is you who runs things, there will be a lot of frustrating times ahead. We

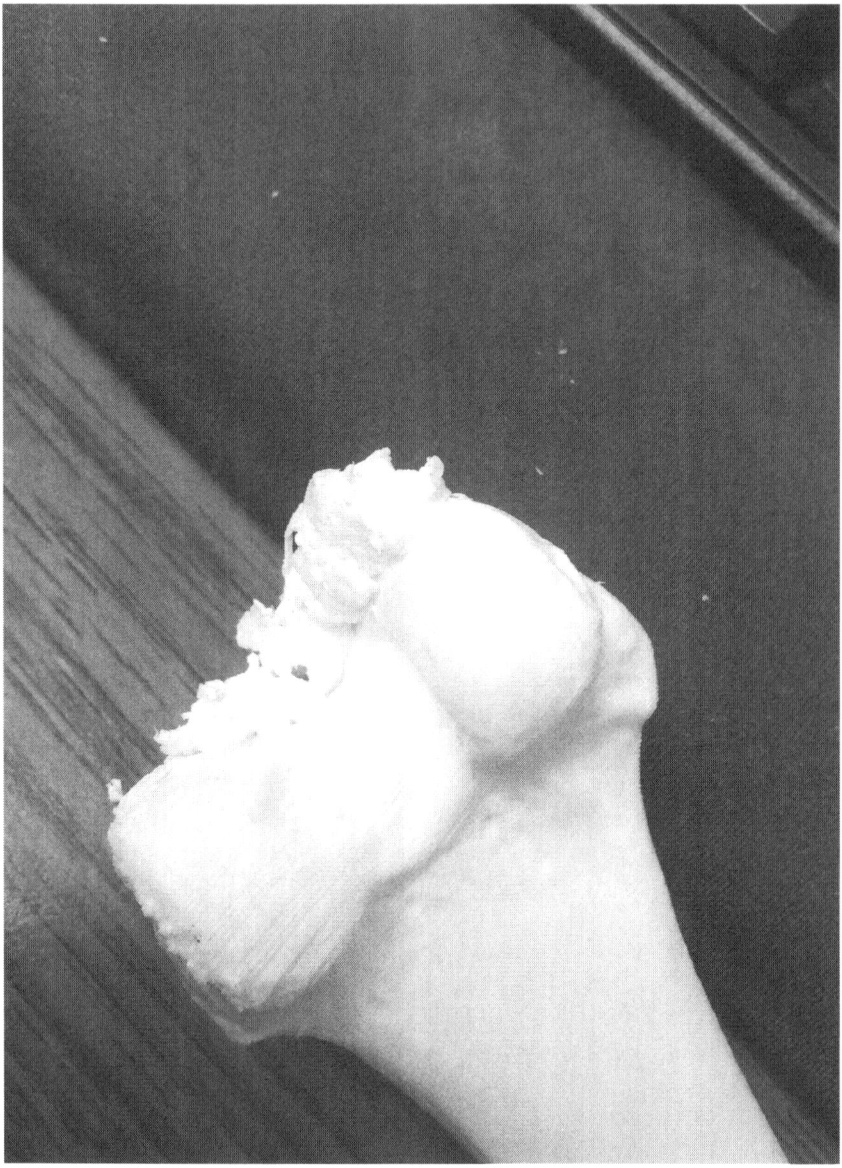

Figure 9.6. A wonderful-looking femur, ruined by support that would not break away. I took this to our 3D printing group and got help—new tool recommendation for postprocessing. A new pair of clippers, and the femur was back to looking good! *Jennifer Herron.*

hired our student employees roughly a year into our service, and even with their help, frustrations can still abound, for both them and you (vicariously). The following is some advice for managing frustrations that will hopefully help on those bad days.

You Will Fail.

Reminiscent of Star Trek's Kobayashi Maru exercise (handling a no-win scenario), 3D printing will test your ability to accept failure. By now you probably understand that 3D printing can be problematic. Print failures are in your future. When starting our 3D printing service at the Ruth Lilly Medical Library, I had no idea that 3D printing had so much that could go wrong. Initially, I thought it was my technology curse that prevents devices from working correctly when I'm around. After impressing the technical support representative for our printer with a malfunction, I thought things were looking grim; however, once meeting with other people managing 3D printing labs and hearing about similar problems, I realized I was entering an incredibly frustrating arena, where nothing is expected to work right. I also learned that this was part of the fun: the experience of 3D printing. As the technology improves, the failures will not occur as often and eventually may no longer be an issue, but in the meantime, it is important to realize that failing is part of the process of 3D printing. After a while, the more I failed, the more I learned and gained the confidence to start more in-depth troubleshooting. Accept failure, and don't let it get you down. The following is some advice for those new to 3D printing or those who may feel like they are failing.

Failing to Plan Is Planning to Fail.

As soon as I planned for a model to be an easy print I typically had some of the worst errors and failings I had ever had in printing. Be careful when telling users when their print will be done; plan on things going wrong and budget some extra time. Filament can run out, extruders can clog, and models can break when being removed from the print bed. Whether it is a printer failure or an operator failure, it will happen when you least expect it. I would rather be surprised that a print succeeded than a print failed.

Know When to Quit.

Like the saying in poker "Know when to hold 'em and when to fold 'em," you should know when the model you are trying to print is a lost cause. It can be hard to give up, but you do not need to quit on the model itself. If there are other print labs on campus, contact them and ask help. After nearly three years of 3D printing, my print lab started budgeting for models to send out to other campus labs for printing. This idea came after a print fail at our lab.

The model managed to fail on our three printers but within a few hours was successfully printed at another lab, and no one understood what went wrong with our printers. This likely might happen with your service, and if it does, it might be a good idea to write in a loose policy on how many times you will attempt to print a model. Based on my review of other libraries with 3D printing services, libraries tend to accept the costs of printer failures, but it is not clear how often they try to print a model. Try to set a limit or gauge of when to know that it's time to move on from a failed model so you can avoid getting caught up in trying to fix what might be an impossible print for your printers (whether it is mechanical issues or actual ability issues).

If You're Going to Fail, Fail Hard.

Don't be afraid of prints that might seem like instant failures. There were many prints that seemed impossible and ended up being the best prints our lab produced. When prints did fail, after sending users an update about the failure, the response, almost always, was appreciation for trying.

Additionally, I also learned failure is in the eye of the beholder. I often kept failed models outside my office that, while they completed printing, had imperfections with support removal, broken sections, or other imperfections. Then one day, a pair of students stopped by and asked if they could take an anatomy model. I showed them the problem with the model and offered to print them a better one, but they liked the failed print because, to them, it showed them anatomy that helped them to understand normal anatomy better. I had similar experiences with broken models that users happily took despite the problems.

Failing Is Frustrating; Find Fail Friends.

Printers fail. Designs fail. Don't take it personally when things fail. Our 3D print group at Indiana University had regular meetings where we discussed projects and potential campus-wide initiatives. Another topic that was just as engaging was failing. It was a fun part of our meetings and helped to reinforce the fact that everyone fails. Having a group to meet with helps to vent frustrations and can provide a support group for troubleshooting.

Have Fun.

Despite all the failing and all the troubleshooting and maintenance involved, 3D printing is a lot of fun—so much so that I am in the process of buying my own printer so that I can keep on printing and failing all the time. While it is easier said than done, don't get too frustrated, and look at every failure as an opportunity to learn what not to do again in the future. Take a break, get some coffee or tea, and try again.

REFERENCES

3D Printing for Beginners. 2018. "How to Store 3D Printing Filament: A Simple and Cheap Way to Keep Your Filament Dry." http://3dprintingforbeginners.com/how-to-store-3d-printing-filament.

Jennings, A. 2018. "2018 3D Printing Troubleshooting Guide: 41 Problems Solved." *All3DP*. July 2, 2018. https://all3dp.com/1/common-3d-printing-problems-troubleshooting-3d-printer-issues.

Landry, T. 2016. "Beat Moisture before It Kills Your 3D Printing Filament." July 28, 2016. MatterHackers. https://www.matterhackers.com/news/filament-and-water.

Ultimaker. 2018. "How to Store Materials." https://ultimaker.com/en/resources/39052-how-to-store-material.

Williams, D. 2018. "Americans Can Legally Download 3-D Printed Guns Starting Next Month." CNN. July 20, 2018. https://www.cnn.com/2018/07/19/us/3d-printed-gun-settlement-trnd/index.html.

Chapter Ten

Basic 3D Model Editing and Printing Prep

Finding a model for 3D printing is only one of the hard parts involved with 3D printing. The other challenge comes from needing to edit models to make them easier to print. This might not be necessary for all models, and depending on the printer you use, the software that creates the printable file might be everything you need. Reviewing terminology will ensure that you have the information you need to make the appropriate adjustments.

BASIC TERMINOLOGY

Infill: the filling of a model. This can be solid, diamond, honeycomb, or lines and determines how solid or hollow your model is.

Layer Height: the width of the layer of filament that will be stacked on top of each other to complete the model.

Orientation: the way the model sits on the surface plane or print bed.

Print Speed: the speed the print head moves as it extrudes and builds up layers for the model.

Resolution: the ultimate clarity of the model after printing is complete. Higher resolution will decrease the layer height and allow for greater detail.

Scale: adjustments to the original size of the model design.

Shells: the outer walls of the model (Digital Engineering, 2016; Redwood, 2016; Yusuf, 2017).

More than likely, models will need some editing prior to printing. Such editing might include reducing the size to fit the print bed and adjusting for a faster print or to print in greater detail. This chapter should help to get you started with the basics and provide some information on adjustments you can make without necessarily having a background in 3D model design.

Scale

First, one of the most basic and possibly most important editing options is the scale of the model. Models created using DICOM might be life-sized and need to be scaled down to fit the print bed as well as reduce print time. Scaling a model can be completed when generating the print file and might allow you to reduce it by entering a new percentage, or it might provide a slider-type tool to increase or decrease the size. Some printer software may also provide an option to print max size or scale to fit for the size of the print bed. Cura is one software program that creates print files for the Ultimaker printer. One nice feature of this software is the ability to scale the model up or down and then see the effect it has on the print time as the software slices the model and provides this information in the lower-right corner.

There is one thing to be aware of when adjusting scale. Typically, printer software includes this option when preparing the print file. However, you are not able to apply this change to the original file. It might seem obvious, but when changing a model's scale, it is important to note the size change and ensure that, if there are other pieces, they receive the same modifications. Additionally, if you print a model at a certain scale and later receive the same request, keeping a record of the scale used will cut down on editing later. Please see chapter 8 for more information about keeping track of data associated with 3D models and printing.

Orientation

The next editing you might encounter is orientation. This is something that can be done automatically with some software, including an autoplace feature. One drawback, however, is different angles that could allow for a faster print or allow a larger model on the print bed might not be considered. For example, if you try to print a model that is longer than it is taller, such as a femur bone, depending on the orientation, the model might not fit on the print bed unless it is rotated to a certain angle. Additionally, there are three axes to take into account: the X, or width; the Y, or length; and the Z, or height. If a model needs to be as big as possible, you can adjust the angle so that the model sits a little higher and is more diagonal on the print bed. Yes, this will most likely require additional support, but this is when you need to determine how important the final size of the model is. On the contrary, if you are looking to save support and are printing something that is somewhat hollow or has openings, such as a rectangular box or a cranium, you can adjust the orientation so that the model sits with the open side up and therefore will not require support. This reduces the print time and saves material and your time in postprocessing the model.

If you want to go a little advanced, there is a software program called Meshmixer by Autodesk that offers model analysis, and without needing any design skills, you can run an auto-analysis that will determine the best orientation for a model for the fastest print time. One comment about this feature: While it is extraordinarily useful, it also has some drawbacks. For instance, I used this when printing a crescent-shaped rectangular piece for a device. While it seemed like the best printing orientation was flat on the print bed, Meshmixer's orientation tool adjusted the model to print at an upward, diagonal angle. It printed faster and with no issues, but when I used this tool again for a skull model, it orientated the skull to be face down. It again printed with no issues and at a faster time; however, the supports that were needed left dimpling after they were removed and reduced the quality of the appearance of the model. When using the orientation tool and when considering orientation for your print, pay attention to where the supports will be and try to determine if this will affect the model's appearance. For more rigid, linear models, this might not be an issue, but for models that have certain features that are more important than others, this is something to take into consideration.

Shells

The number of shells a model has creates a more complex and intricate design. The more complex a model, the larger the file size. This is one area where you might run into problems, depending on your printer software. Some printer software either may not be able to process file sizes or will take a long time to process, during which time it might crash your computer and fail to generate a model. There is a quick and easy way to reduce the number of shells and thereby reduce the file size. Meshmixer offers an editing feature where you can select the entire model and then select the option to reduce shells. One word of caution: When reducing the shells, the model's clarity and intricacy will also be reduced (Redwood, 2016).

Layer Height

Layer height is something that is not so much an edit as it is an adjustment to the print settings. Layer height determines how seamless your final printed model will look. Smaller layer height will make your model look more solid but will take longer to complete. Layer height is based on millimeters. Layers can go up to 3 millimeters and typically to 1 millimeter and, with some more advanced printers, down to 0.2 millimeters. Adjusting the layer height can save time and material, depending on the need and design of the final model.

A proof of concept may be flexible in its layer height, but if you are printing a piece that needs to be detailed and precise, the layer height might need to be smaller (Grames, 2018).

Resolution

Printer resolution is greatly affected by the layer height; however, layer height is only one aspect of resolution (Digital Engineering, 2016).

Print Speed

Print speed is another aspect of printing that can affect how a model turns out. A slower print speed will cause the print time to increase, but it will also allow for greater detail in the final model. Layer height and print speed can both work together or interchangeably to customize the resolution of the model.

Now that the primary editing and modification areas have been described, the software that can be used to complete these changes is described in greater detail next. Additional features and benefits are highlighted for alternate uses of the software.

STAND-ALONE EDITING SOFTWARE

The software described here is outside your printer-specific software, such as MakerBot or Cube. This software can be used to do some basic or more advanced editing depending on your needs and skill level with the software. You may never have to venture into editing models, or you may stick with the 3D printer's software and only do the editing that it allows. If you have the time and the willingness, I recommend testing out editing software to see what more you might be able to provide with your service.

Blender

For more advanced editing, Blender is a good resource that is free to use and made available through a Dutch nonprofit corporation. Blender has a lot of features and goes well beyond basic editing. Learning about the advanced features takes some time but will provide additional services to users interested in prototyping. Blender also provides users with the ability to check models for errors. MeshAnalysis reviews overhangs, thickness, and sharp edges. An additional tool called Print Toolbox identifies advanced errors that may occur with a model. One of the big perks with Blender is the ability to design with precise measurements. Users can input exact angles, and for curved and linear models, the exact length, width, and height can be added (Sculpteo, 2018).

Meshmixer

My go-to software for editing 3D models is Autodesk's Meshmixer. Meshmixer is free to download and can edit models as well as create printable files for some 3D printers. Meshmixer was mentioned previously for reducing the number of shells for large files. One great feature is the analysis option. The analysis feature offers a variety of tools for reviewing and repairing 3D models. An inspector tool reviews the model for imperfections, and you can select to autorepair all, or you can apply individual repairs by clicking the pins highlighting the error. This can be helpful if, after selecting the repair-all option, you notice a change in the model's appearance that takes away an important element. For example, for a heart model I was editing, the repair-all removed some of the smaller veins that were crucial to the final model. By selecting individual pins for repair, I was able to keep the veins and print the model with the other repairs applied.

Another great feature included in the analysis option is orientation. The orientation, as described earlier, determines how the model sits on the print bed. Using the orientation available through the analysis feature analyzes all orientation options and selects the one that will result in the fastest print time. Additionally, another feature that reduces print time is layout/packing. The layout/packing feature is useful if you want to print multiple models at the same time. One important note is that all models in this layout will have to share the same print settings.

If you have any problems with the supports generated by your printer's software, you can test out adding supports through Meshmixer. Supports can be added using the overhangs tool. I leave the settings for the supports at the presets and use these supports as an alternative if I ever have problems with the supports generated by the printer software. One example of this is when I could not get supports to break away cleanly. I used the Meshmixer supports and found that they broke cleaner and enabled faster postprocessing. A drawback with these supports is that they tend to leave dimpling. I recommend only using these if you encounter issues with supports generated by your printer software.

In the edit section of Meshmixer, there are various tools to modify the model design. There is a plane cut tool available that can be very helpful in certain situations, including a model that needs to be printed at a scale larger than the printer bed. Additionally, one trick you can use the plane cut tool for is to get a sectional view of anatomy. This only works with certain models; typically models created with DICOM that were not filled will contain the internal structures. Based on my experiences, students request models with clear exteriors and solid-colored interiors. While our printers could not complete these requests, a compromise was split-model, which was cut in half or segmented in any other way that would display the interior anatomy.

Another editing tool that, again, can be helpful in certain situations is the mirror tool. Mirroring can be helpful if you need to replicate the parallel of a model. I learned about this tool when I created a split-skull model that had a mirrored other half, which then allowed for a whole skull to be created with the two mirrored models.

A final tool that could provide some repair-like editing is the make-solid option. The make-solid tool is just like it sounds: It makes a model solid. This is not something you would necessarily need to do to every model you print, but a model that is hollow and has thin walls may crack after printing or fail to print at all. Making the model solid fills any hollow spaces between the model's walls and prevents any cracking that may occur during postprocessing.

OnShape

If you need some additional editing features but don't want to download and install software, OnShape is a cloud-based alternative. OnShape offers a free subscription with the requirement that models are made publicly available. Boolean operators are available to make combining or separating models easier, and the split feature enables you to print larger models that do not fit on your print bed. An added benefit with OnShape are the workflows that are available to assist in these operations.

Depending on your service, editing models may or may not fall under the library's responsibility. You might also have dedicated staff to manage this aspect of 3D printing. However, if you are looking to offer additional support, this information can guide you through some of the options available. Furthermore, if you have concerns about errors with models, these programs can help to identify and repair models before attempting to print. Again, the technical skills needed for these programs require time and patience to learn but will help expand your service by offering this additional feature.

Tinkercad

If you are interested in doing somewhat more advanced modeling, Tinkercad is another great resource and is completely free, with no software to be downloaded—a perk for those with IT restrictions. Tinkercad is another resource provided through Autodesk. It is cloud-based and only requires an account. Tinkercad does have size limits: Models must be under twenty-five megabytes. Tinkercad also provides a wealth of training and educational resources to help users gain experience in 3D modeling.

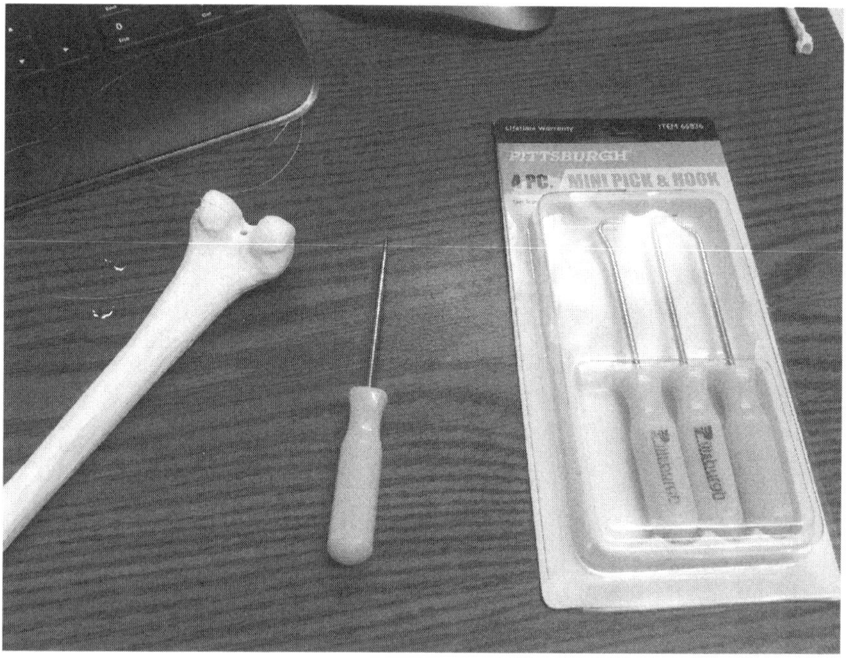

Figure 10.1. While editing models to make holes is easier, there are always alternatives if you are pressed for time. *Jennifer Herron.*

If you are looking to resize a model, Tinkercad is a quick go-to resource. If you also have more creative ideas for 3D printing, Tinkercad allows you to create your own models using basic shapes and text. If you've ever used the group tool when combining elements in a PowerPoint presentation, then you will similarly appreciate the group option in Tinkercad. For printers with dual extruders for multicolor printing capabilities, the group feature in Tinkercad allows for the printer software to recognize the individual models and enable multicolor prints. For the mirrored split skull mentioned earlier, this is where both models can be combined into one model for printing. Any model that might have multiple pieces can have them combined and merged into one model using this tool; another unique way to use this is in conjunction with another feature of Tinkercad—negative models. Negative models are modes that create empty space in a solid model. An example of this is if you need to insert holes in your model. One experimental way I attempted to use this feature was to create a hole in a brain model scaled down to fit inside a model of the skull of a Phineas Gage (see figure 5.7, p. 62). Tinkercad assisted in this project by not only allowing me to scale a separate brain model down to fit in the Phineas Gage model but also allowing me to create a negative model and then match it with the injury shown on the skull model. An alternate way of connecting the two models could be using magnets, which is made easier when adding holes to the models prior to printing. Another method of joining models together is called the jigsaw method, which can be completed using both the group and hole/negative tools.

GO BEYOND THE BASICS: ADVANCED 3D MODELING SKILLS

If you have a need for original designing or prototype assistance, the software here will help make this possible. It definitely requires some time and patience to learn. These programs may also be useful in cleaning up DICOM-generated 3D models or can be used to review models that failed to attempt some repairs.

REFERENCES

Digital Engineering. 2016. "3D Printing Glossary." May 2, 2016. http://www.digitaleng.news/de/3d-printing-glossary.

Figure 10.2. Using Tinkercad's group function to combine models for creative approaches in marketing. *Jennifer Herron.*

Grames, E. 2018. "3D Printer Layer Height: How Much Does It Matter?" *All3DP*. May 28, 2018. https://all3dp.com/2/3d-printer-layer-height-how-much-does-it-matter.

Redwood, B. 2016. "The Definitive 3D Printing Glossary." 3D Hubs. https://www.3dhubs.com/knowledge-base/definitive-3d-printing-glossary.

Sculpteo. 2018. "Blender: Prepare Your Model for 3D Printing." https://www.sculpteo.com/en/tutorial/prepare-your-model-3d-printing-blender/modeling-3d-printing-blender.

Yusuf, B. 2017. "14 Most Important 3D Printing Terms (Glossary of Terminology)." *All3DP*. November 10, 2017. https://all3dp.com/1/3d-printing-terms-terminology-glossary.

Chapter Eleven

Marketing and Outreach

One of the challenges with starting a new service is marketing and promoting it to your intended audience. 3D printing can help to attract people because it's a new and exciting technology; however, ensuring you are reaching your entire target audience and population is always a struggle. Learning about different marketing approaches can help to guide the start or relaunch of a service. Knowing some strategies ahead of the launch of a new 3D printing service, or even learning how to better market an underused service, will help to attract new users. In deciding how to market a 3D printing service, it is best to take a step back and review marketing practices in traditional business, as well as in all types of libraries. By understanding the basics of marketing and how to best market traditional services and products, it will become clearer how to market 3D printing for library users.

MARKETING BEST PRACTICES

The US Small Business Administration (2018) emphasizes the importance of creating a marketing plan that highlights the "competitive advantage" of a service, identifies intended audiences and markets, creates goals, sets a budget, and forms an action plan. Martin Zwilling from *Forbes* also provides some don'ts for marketing specifically aimed at startups. Don'ts include bragging or speaking too highly of the service, making false claims, not following through on promises, networking to audiences that are too large, and not providing enough content or depth to the information shared with customers (Zwilling, 2013). *Business Insider* contributor Sujan Patel discusses the marketing plan itself and the core elements within that create a successful and impactful marketing campaign. Patel identifies six main technology-based skills that benefit marketers: creating engaging videos, launching successful e-mail campaigns, taking advantage of marketing automation services, posting regular and sincere social media posts, and enhancing search engine optimization to make services discoverable. Developing a stand-alone social media marketing plan is also advised (S. Patel, 2015).

When it comes to social media, there are a lot of best practices to make the most productive use of this marketing tool. Some aspects of social media marketing that are not always considered before posting include the timing of posts, tone, length, and tracking posts for effectiveness (Lee, 2016). In addition to a general social media marketing plan, individual platform marketing plans are also encouraged. Hootsuite offers a series of steps to follow when establishing a Facebook marketing plan. Knowing who you are marketing to, creating goals, and following the "social media rule of thirds" are some of the main steps. The social media rule of thirds is a good rule to follow. It breaks down the goals of posts and ensures that posts themselves are not too narrow or too broad. The rule is that one-third of posts should inform, another third should interact with users, and the last third should promote your service. Tracking posts with a spreadsheet is one way to measure how posts rate against this rule, and it can help to plan posts for the future. Using the Facebook business pages is also recommended, as well as using different types of posts—text, photos, video, live video, linked content, and pinned posts (Newberry, 2018). Making sure your social media pages can be found is also important, and suggestions include linking to pages in e-mail signatures. Creating a Facebook group to boost interactions between users is also recommended. Groups can provide a niche area for users to collaborate and comment on the service. Lastly, monthly reviews are also encouraged to see the reactions to posts and determine what works well and what needs to be changed.

A concentrated marketing approach also discusses when marketing services or products is content marketing. Content marketing focuses on making connections between products and services with the consumers or users. For example, while broad marketing efforts can promote 3D printing as a service, a content marketing approach highlights certain aspects of 3D printing and identifies a specific user population and a need that it would fill. For successful content marking, social media posts require developing an authentic voice and creating meaningful posts to speak to these individual groups. A group effort is also recommended for managing social media marketing, as one person can easily become overwhelmed with posting regularly on the various platforms (Qureshi, 2015).

MARKETING FOR LIBRARIES

Public libraries are a good example of organizations that must market to a broad community with a diverse set of users, services, and resources. A 2012 survey found that, when marketing, 67 percent of public libraries focused on mass marketing as opposed to targeted marketing (Dowd, 2013). Nancy Dowd believes these practices led to a "disconnect," where public libraries marketed too broadly without reaching their stakeholders and illustrating their value. Dowd also illustrates a crucial element of marketing that can be missed and forgotten over time: marketing not just the events and services but also the values that these provide users. Telling consumers about a new service versus telling them how the new service benefits them is the key difference between promotion and marketing, as discussed by Laura Solomon (2016). Libraries are great at promoting resources, but when it comes to marketing, they struggle to send the right message. Laurie Russo (2017) discusses content marketing as a solution to reestablish the connection between users and the value of the services libraries provide. Furthermore, along with promoting the value through marketing, it is essential that the messages and information being sent out are tailored to individual user groups so that the message can have more impact. Russo cites another librarian, Danielle Patrick Milam, who facilitates marketing workshops, in the importance of "reaching the right user at the right time." Timing and messages are especially important when it comes to social media.

As described earlier, marketing through social media is frequently brought up for its benefits and ability to broadly share messages to users; however, what seems like an easy tool to use is, in fact, quite tedious and onerous. Numerous articles providing marketing insights and best practices are explicit when it comes to social media—post regularly and thoughtfully. Many authors agree that blogs are a great tool for libraries, but lengthy posts are cumbersome for users wanting a quick bite of information. Social media outlets, such as Facebook and Twitter, provide that quick bite, and others, like Instagram and YouTube, can provide even better, more engaging bites of information for users to consume almost instantly by simply viewing an image or short video. One major no-no, as pointed out by many authors, is sending the same message across all platforms. As Russo (2017) points out, social media has different user cultures. Because of these differences, sending the same message out across these platforms is not going to get the same impact when it is not tailored to the audiences. Understanding the different user groups on different social media platforms will help to determine the message to send.

According to a 2018 Pew study, Facebook is one of the long-standing favorites for social media platforms, but in 2018, YouTube became the most popular platform, with 73 percent of adults using the platform over the 68 percent of adults using Facebook. Instagram came in third, with 35 percent of adults using the platform. WhatsApp and Twitter came in as the least popular platforms, with 22 percent and 24 percent, respectively (Smith and Anderson, 2018). A deeper look at individual platforms also identifies further information about its users. For instance, Facebook, while still being used by 81 percent of users aged 18 to 29, may soon see this number decreasing. In a recent article released by the *Guardian*, three million Facebook users in this age group plan on leaving the service. Parents joining Facebook was cited as the cause of the departure. Facebook is being called "increasingly uncool" with younger adults, who feel that parents and older family members use the platform to monitor them. As a result, younger adults are less keen on posting content (Sweney and De Liz, 2018). Users displeased with the influx of older adults on Facebook have left the platform for Instagram and Snapchat. When asked about Instagram, one user noted how the visual posts are more appealing and relatable than text-filled posts.

With different groups using different social media platforms, it is recommended to tailor messages based on these groups. This is where the time-consuming part of social media marketing comes into effect. Creating custom messages for each platform takes some time, and Solomon (2016) offers some great advice in streamlining this process: Repurpose messages. For example, start with a library blog. A blog post will have more content, and by segmenting that message to fit your other social media channels, you can save yourself from having to create new content for each medium and instead focus on altering the delivery by including quotes from the blog and an image. Furthermore, the frequency of posts with each social media platform also requires regular updates and varies between them. Neil Patel (2016) reviews the frequency for posts and describes the ideal amounts for each platform. One important piece of information shared by Patel is how followers and goals can affect the frequency of posts. University and organization accounts may have thousands of followers, but libraries might only have hundreds. The frequency for posting will be different based on the number of followers. If there are less than ten thousand followers, posts should be less frequent than for accounts with more followers. Study results show that clicks per post can almost double when accounts with smaller followings only post one to five times per month (N. Patel, 2016). With Twitter, Patel found that the number of posts did not matter as much, as a study from Social Bankers found no adverse effects with posting too frequently. However, it is recommended to post one to five times per day to help build engagement with users. Instagram had a similar finding, with no negative effects on posting too frequently; however, it deviates from regular postings, which has a negative effect on followers. A resource guide from the Library Information Technology Association (LITA) echoes these sentiments, adding you also want to ensure that you are not posting too much, as this might be considered spamming to users who do follow your accounts (Thomsett-Scott, 2014). The main thing urged by authors is consistency and quality over frequency.

In addition to timely posts, users also expect timely responses to any questions they may post (Thomsett-Scott, 2014). Leaving questions unanswered not only upsets the person posting it, but also new users may be dissuaded from interacting with the page in the future. Using social media for marketing should not be thought of as a one-way street. Again, this adds to the time and dedication needed for running a successful social media account.

Another trick to social media marketing is gaining followers, and one bit of advice from the LITA guide is to make an effort to find out where the library might be mentioned in user comments or posts. A search can be run using the library's name, or the library itself might be tagged. Liking these posts or engaging or interacting in any other way helps to market the library by illustrating the people behind the page.

In addition to customizing content for each platform, you must ensure that the messages themselves are not "egocentric." As Solomon (2016) describes, content marketing needs to focus on the users and not the library. While it sounds contrary to the goal of marketing, the marketing messages focused on users will reach their users more than messages focused on the library. An example Solomon highlights is that libraries tend to promote the library and an event but fail to make the connection for users about why they should care and how this event benefits them. Solomon refers to this misstep as "youtility" and encourages libraries to advertise for not only their events but also the events in their community. This builds a greater relationship among groups and shows users that the library is more than a place and a space; it is part of their community (Peet, 2016). Finally, Brian Mathews (2009) also encourages libraries to think beyond academics and treat users' needs more holistically. Students come to the library to do more than study; they have other needs, such as charging their devices, getting a bite to eat, or taking a break from the day. Libraries can incorporate these needs into their marketing plans.

After learning more about the best practices of marketing messages and media, the next step is to understand the best practices of organizing your marketing efforts. Having a marketing plan in place is cited as the most important step in marketing and a great way not only to find your audience but also to learn about how to best develop your 3D printing service. The marketing cycle, as discussed by librarian Ned Potter (2012) is an eight-stage process, where libraries establish goals, research their communities, identify specific user groups within their communities, define objectives, begin their marketing endeavor, measure the impact of their marketing efforts, evaluate outcomes, and modify for the next marketing effort.

Once you know what you will be marketing and to whom, the next step is determining how to then push out these messages. As discussed earlier, social media should be used; however, academic librarian Mathews describes a three-week marketing timeline and offers advice on the best time to market on social media and on-site. Three weeks before the event, Mathews suggests putting up posters in the library to advertise the event. During the second week, add posters outside the library, share flyers, and promote on the library website to continue to reach further out into the community. Finally, the week before the event, add table tents and an article in the campus newspaper to drive the event further and keep the event fresh in users' minds (Mathews, 2009).

Throughout this book you may have noticed a trend or common theme that runs through all the elements of establishing a 3D printing service: building relationships. Marketing is no different, and you should continue to build those relationships to not only grow your service but also to create your own user community of marketers and promoters. Word of mouth is still considered a viable marketing strategy, and when marketing, you need to focus on existing relationships and building new relationships with members outside the library who might not have traditionally been considered an outlet for marketing. An example of such relationships can include information technology and informatics groups. In my experience, these two departments were some of the best promoters of the Ruth Lilly Memorial Library 3D printing service and helped us to develop connections with faculty and students who otherwise missed our marketing efforts.

MARKETING PLAN

Knowing what to say and how to say it when dealing with marketing is an important first step. With this knowledge, you can now work on creating a marketing plan that best highlights and features 3D printing. In 2010, Alison Circle shared Linda Hazzan's marketing plan template for libraries to use. Hazzan, a public library director, created a marketing plan that consists of five elements: program goals and objectives, program positioning, target audience and user profiles, marketing and communications program initiatives, and timelines and milestones (Circle, 2010). These elements are then further broken down into three groups to help you to better understand the what, when, and who for the marketing plan. Identifying the goals of the service and how it ties into the overarching goals for the organization is a key step. Additionally, librarians Patricia Fisher, Marseille Pride, and Ellen Miller offer marketing plans with a similar approach in their book *Blueprint for Your Library Marketing Plan* (2006). Making worksheets ahead of time allows libraries to collect data beforehand in order to learn more about their users and community. These worksheets allow libraries to ensure their ideas about their users' interests and preferences are supported by data. Using worksheets to gather data about your users' needs is a great way to prepare for launching a 3D printing service, or for existing services it can help determine if the service needs to change how it is being marketed.

These data worksheets include collecting statistics on the student and faculty population and gathering a current listing of course offerings. In-depth data on faculty and their degrees is also included, as well as rank, research grants, and publication subject areas. Additional data about trends in service and variations in library usage is also on the worksheet. While the data collection for a 3D printing service might not need to be as intensive, it does provide some good ideas of areas to review when considering possible uses of 3D printing for your users. If this data is easy to obtain, the more you can collect, the better the insights you will get about your users; however, it is also possible to collect the data gradually as you build the service. For instance, review the latest research being published by faculty and see if there is any that relates to 3D printing or other forms of innovative technology.

Looking outside libraries is also beneficial when offering a nontraditional service such as 3D printing. The 3D-printing-specific marketing plans I reviewed followed closely with those previously listed, with some minor differences. Differentiating yourself from "competition" was featured in 3D-printing marketing plans and should be considered when thinking about your 3D printing service. If you are located on a university campus and are aware of other print labs, you should consider reviewing your "competition" and seeing how your service stands out (Shapeways, 2018). At my former library, there were at least three other print labs on campus. After roughly two years into our service, the need to distinguish ourselves grew as we needed to justify the continuation of the service. In chapter 9, the service offerings for 3D printing services vary but hold one consistency: models are generally supplied by the users. Medical and health sciences libraries who offer help on finding models have the benefit of using this service to distinguish themselves from other print labs. If libraries look at 3D models as information and market the service in the same way they market literature searches, systematic review services, and databases, it will become easier to understand how to best engage and interact with users.

3D PRINTING SOCIAL MEDIA MARKETING BREAKDOWNS

The following is a summary of social media notes about different message types for different platforms. For all uses of social media, when providing a link to more information, use a trackable link shortener, such as goo.gl, which allows you to track the effectiveness of each platform and message.

Blog Posts

Highlight the benefits of using 3D models for different aspects of user needs. Studying and creating prototyping can be promoted, with more details about how professionals are using these models.

Twitter

Take a snippet from blog, an important quote, or a statistic, and link it to a specific user group. Anatomy students and anatomy faculty are two groups to call out; include a short bit of information applying the benefits of 3D printing for them.

Facebook

Include pictures or videos, with a quick blurb from a blog. If possible, student comments or reviews also help to market the benefits to specific user groups.

Instagram

Stick to just the graphics. A picture of 3D-printed models, a picture of the printer printing, or a picture of users interacting with models all show users the new service. Linking to the blog post or the service home page provides them more details.

YouTube

Creating short videos or commercials about 3D printing draws attention and attracts users to investigate more about the service.

REFERENCES

Circle, A. 2010. "Marketing Plan Templates." *Library Journal*. February 8, 2010. https://lj.libraryjournal.com/2010/02/opinion/bubble-room/marketing-plan-templates.

Dowd, N. 2013. "The Results Are in and They Aren't Good." *Library Journal*. February 5, 2013. https://lj.libraryjournal.com/2013/02/marketing/the-results-are-in-and-they-arent-good-library-marketing.

Fisher, P. H., M. M. Pride, and E. G. Miller. 2006. *Blueprint for Your Library Marketing Plan: A Guide to Help You Survive and Thrive*. Chicago: ALA Editions.

Lee, K. 2016. "For Social Media Beginners and Pros Alike: A Free Social Media Marketing Resource Kit." Buffer. Social Blog. January 29, 2016. https://blog.bufferapp.com/social-media-marketing-resources-kit.

Mathews, B. 2009. *Marketing Today's Academic Library: A Bold New Approach to Communicating with Students*. Chicago: ALA Editions.

Newberry, C. 2018. "The Definitive Facebook Marketing Guide for Business." Hootsuite. March 22, 2018. https://blog.hootsuite.com/facebook-marketing-tips.

Patel, N. 2016. "How Frequently You Should Post on Social Media According to the Pros." *Forbes*. September 12, 2016. https://www.forbes.com/sites/neilpatel/2016/09/12/how-frequently-you-should-post-on-social-media-according-to-the-pros/#e5785b2240f2.

Patel, S. 2015. "Six Marketing Skills Every Business Owner Should Master." *Business Insider*. January 21, 2015. http://www.businessinsider.com/marketing-skills-business-owners-should-master-2015-1.

Peet, L. 2016. "Maximizing the Message." *Library Journal*. September 29, 2016. https://lj.libraryjournal.com/2016/09/marketing/maximizing-the-message-lj-2016-marketer-of-the-year-award.

Potter, N. 2012. *The Library Marketing Toolkit*. London: Facet.

Qureshi, S. 2015. "Why Isn't Content Marketing Working for You?" *HuffPost*. December 10, 2015. https://www.huffingtonpost.com/entry/why-isnt-content-marketin_b_8758348.html.

Russo, L. 2017. "Mastering Marketing: Library Promotion." *Library Journal*. March 9, 2017. https://lj.libraryjournal.com/2017/03/lj-in-print/mastering-marketing-library-promotion.

Shapeways. 2018. "Writing a Marketing Plan for Your 3D Printed Products and Merchandise." https://www.shapeways.com/tutorials/shops/writing-a-marketing-plan-for-your-3d-printed-products-and-merchandise.

Smith, A., and M. Anderson. 2018. "Social Media Use in 2018: Appendix A: Detailed Table." Pew Research Center. March 1, 2018. http://www.pewinternet.org/2018/03/01/social-media-use-2018-appendix-a-detailed-table.

Solomon, L. 2016. *The Librarian's Nitty-Gritty Guide to Content Marketing*. Chicago: ALA Editions.

Sweney, M., and A. De Liz. 2018. "'Parents Killed It': Why Facebook Is Losing Its Teenage Users." *Guardian*. February 16, 2018. https://www.theguardian.com/technology/2018/feb/16/parents-killed-it-facebook-losing-teenage-users.

Thomsett-Scott, B. C. 2014. *Marketing with Social Media: A LITA Guide*. Chicago: ALA TechSource.

US Small Business Administration. 2018. "Marketing and Sales." https://www.sba.gov/business-guide/manage-your-business/marketing-sales.

Zwilling, M. 2013. "6 Key Marketing Do's and Don'ts for Your Startup." *Forbes*. June 18, 2013. https://www.forbes.com/sites/martinzwilling/2013/06/18/6-key-marketing-dos-and-donts-for-your-startup/#18fe5199f527.

Chapter Twelve

The Maker Movement and Maker Health

WHAT IS MAKER CULTURE?

One word you hear frequently with 3D printing is *making*. *Making* is defined just as you would think—the act of making something. Many different organizations and groups define *making* with terms that include *tinkering*, *problem solving*, *creating*, and *inventing* (Core Education, 2018; Good, 2013). *Making*, in fact, serves as an umbrella term that encompasses a variety of activities that show individual acts of making: *hobbyist*, *crafter*, *artist*, and *inventor* (UTeach Maker, n.d.). Making is nothing new to the world and has existed for centuries. Many of the technological advancements of today were made possible through making. As described by *Make:* magazine, making is not new, but technology and tools are now available that make making easier and more accessible (Good, 2013). One of the biggest impacts to making has been the internet. The internet provides a means of communication across the globe, connects makers from all over, and offers many different platforms to use for collaboration. In 2014, a *Time* article listed the number of adult makers in the United States at 135 million (Bajarin, 2014). Since 2014, you can only imagine how much that number has grown, as technology advances every day and provides new opportunities for making.

With 3D printing, you are creating models and objects. It might be a simplified form of making, where you find a model, download, and print; however, keep in mind that printing is still making because you must ensure that the printer is maintained and operating correctly to make the model. 3D printing can be more advanced with making if you design a workshop or lesson around what you 3D-print. For instance, while the Phineas Gage model was a simple form of making to download, it was what was done afterward with the model that shows the true extent of making. Attempting to re-create the injury to the brain was where making happened. Again, this is not too advanced, but it can engage learners and help expand knowledge.

MAKING IN EDUCATION

Schools and educational institutions have taken great interest in making to enhance learning. The idea behind this initiative to promote making in schools follows the belief that making encourages creativity and engages learners in more ways than standard forms of teaching may involve (Martinez & Stager, 2014). When making, students become active learners and can focus on how they learn best in their making. Tactile learnings can do the physical aspects of making and put things together. Visual learners can focus on the design elements. Verbal learners can discuss their ideas for making with others, and auditory learners can benefit from this discussion, which supplements information their teacher has already shared.

Furthermore, making teaches additional skills that are not quite as obvious to the learners—teamwork and communication. Making is something that can be done solo, but frequently people come together when making to discuss ideas, troubleshoot problems, and get perspective on their creations (Hertz, 2012). Making helps learners to realize the benefits of collaboration and working in a group, and these are skills that they will take with them long into their lives. Robbie Torney (2015) highlights the importance of making in education in an article published online through the Teaching Channel. Torney states, "We need Making in education to prepare students for a world that is increasingly global, increasingly technological, and increasingly complex." This need for making goes beyond K–12 education and can easily be applicable for all educational environments. Examples of making in schools range from simple project ideas to more thought-provoking activities. The Maker Education Initiative, or MakerEd, provides numerous resources for educators, from lesson plans to professional development to a tool kit for creating making spaces.

MAKER CULTURE

William Craig (2015) discusses the openness and "spirit of sharing" associated with making and the necessity of sharing when describing makers and the maker movement. Without this spirit of sharing, Craig believes the movement would not continue to survive. When describing making, it is hard not to mention the people and their actions and values; these values are referred to as the maker culture, and understanding this culture will help to develop your 3D printing service and plan for activities and events.

In a 2015 article from *The Atlantic*, John Tierney describes potential pitfalls of the maker movement and culture and four main challenges. The top challenge: education. Tierney highlights that the skills needed for making range from basic to advanced and feels that educational shortcomings could hamper the movement. Luckily, collaboration between makers can help with any knowledge gaps and provide lateral learning opportunities for makers.

Legal issues are another problem that could sideline the maker movement and involve issues described in chapter 2, such as intellectual property (IP). Again, the law has tried to address these issues as they arise, but technology has made it increasingly easy to violate IP laws. Some makers might do this intentionally, whereas others might do it without realizing that they are infringing on others' rights. This is a learning opportunity where maker groups can reach out to the legal community for education and guidance.

Another challenge Tierney addresses is the inadequacy of access to the technologies involved with making. As a librarian, the answer is quite clear—libraries! Libraries already offer equal access to information and technology, and adding makerspaces is already something that libraries have started to do to support to equal access for their communities. This, however, is only one outlet for this technology, which is also why many educators are pushing for schools to include makerspaces as well.

Finally, the last challenge seen with making includes moral, ethical, and criminal activities. A popular sentiment from the *Spiderman* comic highlights this last point well: With great power comes great responsibility. Technology has advanced greatly over the years and made possible lifesaving human organs to be printed as well as life-ending weapons to be printed. The law is struggling to keep up with technology, and because this technology is becoming more accessible to everyday users, it opens the door to many misuses. Makers themselves must be responsible for what they create.

Now that you understand making and maker culture more, the next step is finding makers in your community. One easy way to do this is to hold an event. Using making buzzwords like *3D printing* will catch the attention of the right people and pull some from the woodwork. I still remember passing out flyers for my library's MakerHealth Faire and the awestruck reaction from one faculty member when he saw the event's subject—he had no idea that there were others 3D-printing on campus and was excited for the opportunity to collaborate. Additionally, events like this attract the attention of users who are curious about the subject but feel they lack the skills or knowledge to create anything. This is where the maker mind-set comes into play.

MAKER MIND-SET AND DESIGN THINKING

The maker mind-set is what allows for making to take place. As Dale Dougherty (n.d.), creator of *Make:* magazine says, "We can create a workshop or makerspace, and we can acquire tools and materials, but we will not have succeeded at creating innovative thinkers and doers unless we are able to foster a maker mindset" (2). Bringing 3D printers into the library will only be successful if users know what they can do and how they can take advantage of the technology. Developing and promoting maker mind-sets will in turn help to promote the 3D printing service as users become more inclined to make. The maker mind-set is finding its way into education curriculums from elementary school to the university level.

In universities, the maker mind-set is being promoted with a different phrase—*design thinking*. Design thinking shares a similar definition as *maker mind-set*; according to the Interaction Design Foundation, *design thinking* is when people learn to "challenge assumptions and redefine problems in an attempt to identify alternative strategies and solutions that might not be instantly apparent with our initial level of understanding" (Dam and Siang, 2018). Design thinking involves a five-step process for reviewing problems and finding solutions: identifying the problem, applying knowledge learned to the problem, thinking of possible answers, developing a tangible solution, and testing and building on that solution (Sniukas, 2015). While maker mind-set courses are not readily available on college and university campuses, design thinking courses and certifications are available. Stanford, Harvard, and MIT are just some of the well-known universities offering courses and specializations in design thinking. Furthermore, massive open online courses (MOOCs) also offer numerous design thinking opportunities and can, for the most part, be taken freely.

Design thinking has also gained traction in medical education. Curriculum designers are looking into integrating design thinking in the development of curriculums themselves, and the "human-centered" focus of design thinking makes it a fitting subject to be included in coursework (Anderson, 2017; Ku, Shah, & Rosen, 2016). Stanford Medicine also has integrated design thinking into a new summer event, the Design Thinking Medicine Challenge, where students come together to learn about design thinking, apply its principles immediately by identifying health issues in their community, and designing solutions based off collaboration between their peers (Stanford Medicine, 2018). Thomas Jefferson University took design thinking a step further and created the Health Design Lab, "where Health and Design collide in the most exciting ways" (Health Design Lab, n.d.). The Health Design Lab supports students and faculty as they explore the possibilities of design thinking applications in medicine. Most recently, a Health Design Bootcamp (2018) was held over the course of two days. Students and professionals who attended the sold-out boot camp worked together on teams to create inventive solutions to some of the health-care challenges of today.

PUTTING IT TOGETHER: MAKERSPACES

The Health Design Lab is one example of the next step needed once maker mind-sets and design thinking are encouraged—a space in which to begin working. Traditionally called makerspaces, these spaces provide tools and equipment to support making efforts and provide a dedicated space for interaction and collaboration between makers. Other names that are sometimes used to describe similar spaces are fab labs (fabrication labs), techshops, and hackerspaces. These names may seem like alternate names for makerspaces, but they are actually quite different. Fab labs focus on digital fabrication only, and techshops are makerspaces that require paid memberships. Hackerspaces, according to Makerspaces.com (2018), are what led to the creation of makerspaces, as hackerpaces "hack" technology and objects to make them do something other than their designed purposes. Makerspaces follow a similar notion, but rather than alter original functions, more original creation is involved.

Now it is time to get into the nitty-gritty of what *makes* a makerspace. There are no real set guidelines that state you must have certain equipment or tools available. A review of various makerspaces online shows that equipment varies greatly; from button makers and sewing machines to laser cutters and 3D printers, how the makerspace is furnished is typically dependent on the user community's needs (Fallows, 2016). These spaces can provide opportunities for makers to work or offer a space for educational events that provide instruction and guidance in making. Makerspaces address the concern described by Tierney (2015) by providing open access to the technology and resources.

In 2012, President Obama created the Maker Education Initiative to support educational efforts in making and innovation (Kalil, 2015). Two years later, on June 18, 2014, the White House hosted a White House Maker Faire. It was the first step in a series of events that President Obama put forth to promote making in the United States. Additional events include a national week of making, usually around the third week of June, and a Mayor's Maker Challenge (White House Office of Press Secretary, 2016; Kalil & Patel, 2014; Obama White House, 2018). The Institute of Museum and Library Services has regularly participated in the Capitol Hill Maker Faire and complements the efforts of the White House's Maker Education Initiative (Institute of Museum and Library Service, n.d.).

Schools, universities, and libraries are primary locations where makerspaces have opened due to the free access and climate for learning. Teachers are pushing for a makerspace to be available in every school, with some going so far as wanting every classroom to offer makerspace elements. Grants are available to help support this initiative, and MakerBot is also donating to the cause and providing resources and training to support making in classrooms (Millstein, 2013). Libraries have taken advantage of technology grants to install makerspaces, and student technology funds, as well as donations provided by friends of the library groups, have also allowed for makerspaces to flourish inside libraries.

MAKERHEALTH

The maker mind-set took over Jose Gomez-Marquez and Anna Young when they founded MakerHealth, a specialized, more focused area of making with the intentions of turning "providers into prototypers" (Nuwer, 2016). Founded in 2008, MakerHealth focuses on finding creative solutions to problems in the health and medical field (MakerHealth, 2016a). The organization's website provides resources to support those early in the making process with making hardware, software, and electronics. Recommended readings are also provided to help members keep up to date on the latest in the field (MakerHealth, 2016b). After creating a free account, users can see what others have been working on in the "Create" section and upload their own creations. Five years after MakerHealth was founded, in 2013, an offshoot formed called MakerNurse. MakerNurse has the same goal of MakerHealth with a narrower focus on nurses. Workshop requests are available through an online form, and a blog is available for promoting interaction and collaboration.

MakerHealth soon grew beyond a mind-set, and MakerHealth spaces began opening. David Marshall and Deborah McGrew (2017) describe the need for such spaces, as nurses and health-care workers have developed theories to solve problems they encountered but had no way to bring these ideas to fruition. Out of this need, the first MakerHealth space opened in the John Sealy Hospital at the University of Texas Medical Branch (UTMB). Launched in 2015, the MakerHealth space at the UTMB was the first in a hospital setting. The design of the MakerHealth space offers tools and supplies for on-the-spot making, including the basics—tape, glue, and other fasteners—to the more advanced—laser cutters and 3D printers. Additionally, a "selfie station" is in place for users to get pictures of their creations. Being in a hospital, it might seem tempting for users to take their prototypes into the clinical setting for testing, but a quality improvement review board inspects devices made in the space before they can enter any patient care setting (UTMB Health, 2015).

Bringing 3D printers into the library is just one step in the process of developing a successful service. Understanding the elements that come together to make users use the printers is the best way to support the service and foster its growth. When considering a 3D printing service, it is similarly important to consider making and the maker mind-set or design thinking. Encouraging these behaviors will lead to more successful use of the service and urge others to explore making and putting the library's 3D printers to good use.

REFERENCES

Anderson, J., C. F. Calahan, and H. Gooding. 2017. "Applying Design Thinking to Curriculum Reform." *Academic Medicine* 92, no. 4: 427. DOI: 10.1097/acm.0000000000001589.

Bajarin, T. 2014. "Why the Maker Movement Is Important to America's Future." *Time*. May 19, 2014. http://time.com/104210/maker-faire-maker-movement.

Core Education. 2018. "Ten Trends 2015." https://core-ed.org/research-and-innovation/ten-trends/2015/maker-culture.

Craig, W. 2015. "What Is Maker Culture and How Can You Put It to Work?" *Forbes*. February 27, 2015. https://www.forbes.com/sites/williamcraig/2015/02/27/what-is-maker-culture-and-how-can-you-put-it-to-work.

Dam, R., and T. Siang. 2018. "What Is Design Thinking and Why Is It So Popular?" Interaction Design Foundation. https://www.interaction-design.org/literature/article/what-is-design-thinking-and-why-is-it-so-popular.

Dougherty, D. n.d. "What Is the Maker Mindset?" Accessed May 30, 2018. https://llk.media.mit.edu/courses/readings/maker-mindset.pdf.

Fallows, D. 2016. "How Libraries Are Becoming Modern Makerspaces." *The Atlantic*. March 11, 2016. https://www.theatlantic.com/technology/archive/2016/03/everyone-is-a-maker/473286.

Good, T. 2013. "What Is 'Making'?" *Make:*. January 28, 2013. https://makezine.com/2013/01/28/what-is-making.

Health Design Bootcamp. 2018. "The Bootcamp." https://healthdesignbootcamp.com/program-info.

Health Design Lab. n.d. Accessed September 26, 2018. http://design-health.com/hdl.

Hertz, M. B. 2012. "Creating Makerspaces in Schools." Edutopia. November 6, 2012. https://www.edutopia.org/blog/creating-makerspaces-in-schools-mary-beth-hertz.

Institute of Museum and Library Services. n.d. "Makerspaces." Accessed May 30, 2018. https://www.imls.gov/issues/national-issues/makerspaces.

Kalil, T. 2015. "A New Resource for Bringng Making into Education." White House, President Barack Obama. January 12, 2015. https://obamawhitehouse.archives.gov/blog/2015/01/12/new-resource-bringing-making-education.

Kalil, T., and R. Patel. 2014. "Challenging Mayors to Help Make a Difference." White House, President Barack Obama. May 15, 2014. https://obamawhitehouse.archives.gov/blog/2014/05/15/challenging-mayors-help-make-difference.

Ku, B., A. Shah, and P. Rosen. 2016. "Making Design Thinking a Part of Medical Education." NEJM Catalyst. June 30, 2016. https://catalyst.nejm.org/making-design-thinking-part-medical-education.

MakerHealth. 2016a. "Our Story." http://www.makerhealth.co/about/story.

———. 2016b. "Resources." http://www.makerhealth.co/learn.

Makerspaces.com. 2018. "What Is a Makerspace?" https://www.makerspaces.com/what-is-a-makerspace.

Marshall, D. R., and D. A. McGrew. 2017. "Creativity and Innovation in Health Care: Opening a Hospital Makerspace." *Nurse Leader* 15, no. 1: 56–58. https://doi.org/10.1016/j.mnl.2016.10.002.

Martinez, S., and G. Stager. 2014. "The Maker Movement: A Learning Revolution." International Society for Technology in Education. July 21, 2014. https://www.iste.org/explore/articledetail?articleid=106.

Millstein, B. 2013. "MakerBot Academy." MakerBot. October 22, 2013. https://www.makerbot.com/stories/news/announcement-makerbot-academy-and-americas-classrooms.

Nuwer, R. 2016. "The First Makerspace in a Hospital." *Popular Science*. April 14, 2016. https://www.popsci.com/first-makerspace-in-hospital.

Obama White House. 2018. "Nation of Makers." https://obamawhitehouse.archives.gov/node/316486.

Sniukas, M. 2015. "Design Thinking + Business Model Innovation." InnovationManagement.se. August 25, 2015. http://www.innovationmanagement.se/2015/08/25/design-thinking-business-model-innovation.

Stanford Medicine. 2018. "The Design Thinking Medicine Challenge." https://med.stanford.edu/coe/pre-med-students/DTMedicine.html.

Tierney, J. 2015. "The Dilemmas of Maker Culture." *The Atlantic*. April 20, 2015. https://www.theatlantic.com/technology/archive/2015/04/the-dilemmas-of-maker-culture/390891.

Torney, R. 2015. "What Is Making?" October 30, 2015. Teaching Channel. https://www.teachingchannel.org/blog/2015/10/30/what-is-making.

UTeach Maker. n.d. "What Is Making?" Accessed May 21, 2018. https://maker.uteach.utexas.edu/making.

UTMB [University of Texas Medical Branch] Health. 2015. "Nation's First Medical Makerspace Opens inside Texas Hospital." September 25, 2015. https://www.utmb.edu/newsroom/article10648.aspx.

White House Office of Press Secretary. 2016. "New Commitments in Support of the President's Nation of Makers Initiative to Kick Off 2016 National Week of Making." June 17, 2016. http://www.weekofmaking.org/wp-content/uploads/2016/03/2016-National-Week-of-Making-Fact-Sheet.pdf.

Chapter Thirteen

From the Experts

3D Printing in Medical Libraries

This chapter presents opinions from a variety of health and medical professionals. These come from previous users of the Ruth Lilly Medical Library's 3D printing service, and others were solicited from professionals in social media for their medical innovations. All respondents were e-mailed and asked to give their honest feedback on what they thought about a 3D printing service available for students, faculty, and staff in medical libraries. Almost all who were contacted replied and were enthusiastic about the subject. These responses include professionals from different specialties to provide a well-rounded outlook on the service. While these individuals were already inclined toward technology and innovation, it is their perspectives and experience in the field that should provide libraries insights into the potential of 3D printing and their thoughts about whether libraries should partake in this service.

THE EXPERTS

Responses include faculty members, staff, clinicians, and a medical student. Backgrounds include radiology, family medicine, occupational therapy, pathology, surgery, pediatrics, and pharmacy.

Nicholas Anton, MS

Surgical Skills Coach, Department of Surgery, Indiana University School of Medicine

Anton has assisted in research and helped publish numerous articles on the topic of assessing surgical skills and developing a mental skills curriculum (Stefanidis et al., 2017). Recently, Anton helped write an article on how simulation can be incorporated into the "lifecycle" of surgeons as they move from students to professionals, which was featured in the American College of Surgeons online publication *Rise (Resources in Surgical Education)* in early 2018 (Perone, Anton, and Gardner).

On 3D printing in medical libraries, Anton describes the impact the Ruth Lilly Medical Library's service had on his work:

> The 3D printing service at Ruth Lilly Medical Library has enabled our Surgical Education Research Team to engage in research projects that were previously unachievable due to logistical or financial constraints. 3D printing has effectively allowed us to construct an EEG-based headset that would have cost us thousands of dollars to purchase at a significantly more reasonable cost. This headset will enable our team to research elements of surgical skill acquisition that have not been previously explored.

Rafia Rodney L. Duvra

Surgical Skills Education Assistant to Lisa Fisher, Research and Training Coordinator at the Surgical Skills Training Center, and Research Assistant to Dr. Dimitrios Stefanidis, MD, PhD, Chief of Minimally Invasive and Bariatric Surgery and Vice Chair of Education, Department of Surgery, Indiana University School of Medicine

The Ruth Lilly Medical Library worked with Duvra and his team to create simulation models for advanced laparoscopic skills. Duvra also collaborated and provided his insights on new 3D-printing technology using elastomeric materials and molds for surgical simulation of central line placement models. When asked about the 3D printing service at the library, Duvra describes how the service helped create models for surgical simulation:

> With the help of the Lilly library 3D print lab, we've been able to have tasks for advanced laparoscopic skills (ALS) made. I've attached a photo [figure 13.1] of task comparisons with the new and old models. Hopefully, when the time comes, these new models will hold up to the challenge.

Brian Overshiner, BS RT(T)

Senior Radiation Therapist, CT Simulation and Treatment Coordinator, and Volunteer Clinical Lecturer in Radiation Therapy Technology, Department of Radiation Oncology, Indiana University School of Medicine, Indiana University Simon Cancer Center

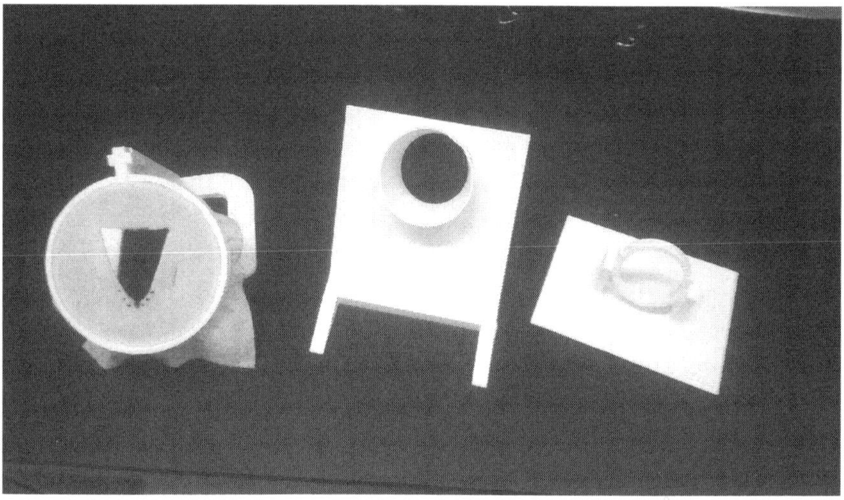

Figure 13.1. Models to support advanced laparoscopic skills. ***Rafia Rodney L. Duvra, Indiana University School of Medicine.***

Overshiner is a member of the Collaborative Additive Manufacturing Research Initiative (CAMRI) at Indiana University. CAMRI brings together faculty, staff, and students from across the campus with subject areas including mechanical engineering, radiology, information technology, informatics, and engineering technology. Projects the group has worked on include process optimization, reducing production costs, printing ceramics, and printing prosthetics and orthoses (Indiana University–Purdue University Indianapolis, 2018b).

Overshiner's most recent endeavors include printing patient-specific boluses for radiation therapy. Partnering with mechanical engineering, his group hopes that they will be able to make the Simon Cancer Center the first cancer treatment facility using this technology for patient care in the country. Additionally, future efforts involve patient education and brachytherapy. A recent grant will provide Overshiner with more opportunities to explore purchasing higher-quality printers for faster and more precise printing (Purdue School of Engineering and Technology, 2017):

> I think medical libraries are a great place to centralize a printing service. Especially in the case of the medical library here at IU, it's a great location, high-traffic area where advertising such a space/service I think would do great. Not to mention you would expect all the residents to migrate through the library at some point in the studies. I think access and advertising is key to making a space work.

I've found that not a lot of MDs realize what 3DP applications can do for them. Teaching about 3DP medical applications could also be done through the library space. Getting young doctors to be comfortable working with digital programs, etc., will help more of them adopt the technology in practice. It is a high learning curve to be able to manipulate and segment models with all the various programs out there. It can be frustrating and very time-consuming for someone to learn (myself included).

Dr. Wallach here at IU is very interested in putting 3DP into curriculum and I could easily see that being taught at the library. Obviously, space and budget is always a concern. I suspect many of these projects will be funded through a grant mechanism or the university, and then finding the right people to run the lab and work with the MDs. I'll think about this some more, but overall, I would see the library as being the starting point to learn 3DP basics and teach classes. Once that knowledge is gained and MDs want to put it into clinic use, I think it makes sense to have an in-house lab or manufacturing center within the hospital.

Joyce M. Lee, MD, MPH

Robert P. Kelch, MD Research Professor of Pediatrics and Pediatric Translational Researcher in Pediatric Endocrinology, Department of Pediatrics and Communicable Diseases, Michigan Medicine, Child Health Evaluation and Research (CHEAR) Unit

Dr. Lee's research interests include health applications of emerging technologies, creating learning health systems that include patient-centered design, the maker movement, and collaborative innovative networks. Additionally, Dr. Lee is the colead of Health Design by Us, an interdisciplinary collaborative that focuses on involving both patients and physicians in the maker movement (University of Michigan, 2018a, 2018b).

Dr. Lee has presented on numerous topics in making and design thinking, including TEDx Detroit Talk, "Participatory Design and the Making of Health"; Edge Talk for the Horizons Group of NHS England, "The Maker Movement: A Model for Healthcare Transformation?"; Cusp Conference, Chicago; Participatory Design, "The Making of Health"; and the Future of Diabetes Technology, "The Intersection of Design, the Maker Movement, and Healthcare" and "Social Media for Patients and Healthcare Professionals." Slides and recordings, as well as workshop guides and other informative resources on design thinking, digital citizenship, and life hacking, are available online at Dr. Lee's website (Lee, 2018).

Dr. Lee's feedback touches on part of Overshiner's response, in that it can be difficult to see the possible uses of 3D printing and its applications:

I would say that even as someone who embraces the maker movement, it is hard to imagine myself going out to 3D print someone because it feels like there is a high bar for expertise needed to be able to design anything. It starts with even identifying the need to 3d print (my patients have diabetes so they need more digital solutions rather than physical ones). I have heard of different programs but have yet to dive in and figure out which one is the most user friendly. Furthermore even if I built one prototype it seems like scaling the invention would be a challenge as well. One other consideration is that many of the "innovations" might be more patient facing so I don't know what solutions need to be generated. For example, my patients may want an insulin pump case or another device carrier, but they know better about the optimal design since they carry and wear the device but I don't. So it's patients that need to be engaged not as much the provider in those cases.

Jonathan Merrell, MD

Assistant Professor of Clinical Pediatrics, Department of Pediatrics, Indiana University School of Medicine; Director of Clinical Care Innovation Accelerator, Indiana Clinical and Translational Sciences Institute; and Pediatrician, Indiana University Health System

Dr. Merrell collaborates with other faculty and researchers at Indiana University (IU) to help transform their ideas into reality. Dr. Merrell has given numerous lectures at IU's School of Medicine to students, faculty, and clinicians. He presented a Grand Rounds topic, "Innovation U. Patents, Brainstorming, and Other Keys to Thinking Like a Biomedical Innovator," and "Anyone Can Innovate!" at the 2018 IU Health Technology Symposium. Information is available in chapter 14 to view Dr. Merrell's presentations.

In a 2017 interview, when asked about the lack of clinician participation in innovation, Dr. Merrell found three primary reasons: time, uncertainty of where to begin, and lack of technical skills. Dr. Merrell is an advocate for innovation and said, on fostering innovation, "The first step of innovation is not to find the perfect solution. . . . The first step is to find the perfect problem" (Zeek, 2017).

When asked about a 3D printing service in medical libraries, Dr. Merrell provided a good insight into the role libraries can play in fostering this innovation:

> It's obvious to me that 3D printing is going to become ubiquitous. Perhaps someday it will become as commonplace as 2D printing. 3D printing is such a powerful platform, and we have only started to scratch the proverbial surface for what can be accomplished through it (not only in the sciences but in art, architecture, education, and other fields). The sooner we can put 3D printing into the hands of designers everywhere the sooner we will realize its true potential. I think that it's a natural next step to make this technology accessible to more people, and I think libraries are an excellent mechanism of achieving that objective. Someday the average elementary school will have a 3D printer in every classroom. It's a logical fit for libraries to guide us in the interim.

Robert S. Pugliese, PharmD, BCPS

Emergency Medicine Clinical Pharmacist, Department of Pharmacy; Clinical Assistant Professor, College of Pharmacy; Associate Director, JeffDesign, The Sidney Kimmel Medical College; and Founding Director, Health Design Lab, Thomas Jefferson University and Hospital

I had the opportunity to speak with Dr. Pugliese about his thoughts on 3D printing in medical libraries. Dr. Pugliese described his current work at the Sidney Kimmel Medical College with 3D printing and the growing interest in 3D printing (Medicine + Design, 2018). He noted how 3D printing is spread across their campuses and living all over, which, he added, is why it should exist in the library—because that's where you find the students and teachers. He believes that, if a university does not already have a place dedicated to innovation, libraries can fill this void by providing the place and space to support these initiatives. He feels that libraries democratize access to knowledge, information, and technology and that 3D-printing technology should be accessible to all students and not limited to department use only. Libraries provide the opportunity for students and faculty to become comfortable with and get their hands on the actual technology and demystify how the machines function.

Dr. Pugliese also provided insights on setting up a 3D printing service by describing how the space at Jefferson University, JeffDesign, was created. Understanding the users is the most important factor to consider, and identifying their needs and growing your service around use cases is recommended. Also recommended is including subject-matter experts in the service. Doing so will help offer more than just a service and enable actual, usable, functional models to leave the space. Furthermore, Dr. Pugliese stresses that starting the service does not require expensive, sophisticated machines. When starting their space, Dr. Pugliese said they started with a single Ultimaker printer and focused on teaching users to 3D-print; as the demand grew, the service expanded.

Dr. Pugliese also provided a great example of how perception of the service can change once use cases are developed. He described a group of ear, nose, and throat surgeons who at first did not understand the need for the 3D printing service. After six months of seeing the use of models for surgery, they now complain if they do not get 3D-printed models to review. This example highlights one of Dr. Pugliese's recommendations on building the service around users' needs. Understanding the needs of users before getting too far ahead with the service will help to develop a more successful service. Furthermore, Dr. Pugliese added, thinking about the human factor is important, regarding not only the user needs but also the needs of those running the service. Time is a serious dedication, as well as learning how to operate the machines and software.

Libraries, Dr. Pugliese believes, follow the trends and offer modernizing service areas that provide access to new and late-breaking technology. As such, he supports medical libraries offering a 3D printing service. Dr. Pugliese provided a wonderful discussion on 3D printing and offered great insights into how to best establish a 3D printing service and how to engage users. He is a great example of how faculty can bring innovation into the classroom and get students thinking outside the box with new technologies.

J. Robin Janson, OTD, OTR, CHT

Clinical Assistant Professor, Indiana University, Department of Occupational Therapy, School of Health and Rehabilitation Sciences

Dr. Janson teaches occupational therapy with a focus on hand and upper-extremity rehabilitation. Dr. Janson works with OT students and provides hands-on lessons in 3D printing and printing assistive devices for patients. She has previously spoken at the Indiana University MakerHealth Faire on designing assistive devices with her students and continues to present on the capabilities of 3D printing in health and rehabilitation.

Most recently, Dr. Janson visited the University of South Dakota and presented two workshops to students and practicing OT professionals. The "Introduction to Digital Design and Fabrication (3D Printing) Workshop for OT Practitioners" provided an overview of 3D printing and how it can be applied in occupational therapy work. She taught attendees how to create models, test safety and functionality of models, and use the 3D printers themselves (University of South Dakota, 2018). Dr. Janson plans to continue holding similar workshops at Indiana University to engage students and faculty locally as well as visitors. When asked about 3D printing in medical libraries, Dr. Janson provided a great perspective and helped to identify the gap in skills that librarians might encounter, as well as how to make a well-rounded service that will be most effective and successful:

> I think it takes a librarian who also has a strong skill set in digital design and fabrication to serve in the position well (likely not very many) . . . or have someone without a library degree who has the skill set work within the library (maybe an easier find). Also it is important that this person be connected so that he/she can access 3D printing equipment/services beyond the capabilities of what the library can offer.
>
> Having 3D printing services available for students and faculty is great (although without students/faculty having much experience I wonder if they often have unrealistic expectations as to what can be printed). I think it would be great for the libraries to offer educational workshops on 3D printing (back-

ground, how to acquire the digital designs, print set up, 3D printer operation, post-print processing, etc.).

Shannon Cooper, MSEd

Instructional Design Manager, Department of Family Medicine, Indiana University School of Medicine

In 2016, Cooper joined the Interprofessional Practice and Education Center (2018); she serves on the TEACH! Curriculum Design Team and participates in the Association of American Medical Colleges Subgroup on Information Resources (Indiana University, n.d.).

In 2016, Cooper, Scott Renshaw, and Zebulun Wood were awarded a $10,000 Curriculum Enhancement Grant through the Indiana University Center for Teaching and Learning for their project "Enhancing Learning through Augmented Reality with 3D Print Technology." Renshaw, assistant professor of clinical family medicine, and Wood, lecturer and codirector of Indiana University's media arts and science undergraduate program, worked together with Cooper in the creation of a learning aid for third-year medical students and graduate media arts and sciences students (Indiana University–Purdue University Indianapolis, 2018a). The learning aid was a 3D-printed knee joint model that not only provided students hands-on learning but also acted as a trigger for an augmented-reality model for further interaction. In her response, Cooper highlights the gap that medical libraries could fill by offering a 3D printing service.

> 3D print labs in medical libraries bring resources to both medical students and faculty that would most likely not be available otherwise. The 3D-printed models are an inexpensive way for students to see and feel as they learn. Anatomy 3D prints have been most valuable in having a low-cost, accessible way to teach about musculoskeletal exams on the various joints.

Michael Bartellas, MSc in Med (AHSR), BSc (Hons)

Doctor of Medicine Program, Class of 2019, Faculty of Medicine, Memorial University of Newfoundland

Michael Bartellas, a fourth-year medical student with Memorial University of Newfoundland (MUN), recognized a need for 3D printing when he and fellow medical student Stephen Ryan were unable to access the technology through local resources. Using their own money, they each purchased their own printers and reached out to faculty and students, offering their 3D printers for use. Soon after, Bartellas began working with an orthopedic surgeon to 3D-print heel fractures in a retrospective study and for medical education purposes. After the success of this work, Bartellas and Ryan purchased new printers: an Ultimaker 2+ and LulzBot. Along with the printers, the pair also received space in MUN hospital, where they founded MUN MED 3D (Memorial University, 2018a). As the service continued to grow, so did Bartellas's aspirations for 3D printing in medicine. Along with promoting 3D printing at MUN, Bartellas wants MUN MED 3D to serve as a launching pad to offer this service remotely to rural health-care professionals and patients. Currently, Bartellas is participating in ongoing projects within MUN MED 3D that include simulation, education, preoperative evaluation, and prostheses (Memorial University, 2018b). As a result of his work, Bartellas was the recipient of the 2017 Charles Tabachnick Canadian Medical Hall of Fame Award for leadership and innovation in health care (Osmond, 2017; Scott, 2018).

After coming across Bartellas's article on setting up MUN MED 3D, I e-mailed him for his feedback about offering a 3D printing service in medical libraries. He graciously gave me the opportunity to speak with him directly on the topic. Bartellas offered great feedback on 3D printing in medical libraries and confirmed that medical libraries offer the sustainability needed to allow for medical students to engage with 3D printing. With the library managing the requests and printer operations, students and faculty can experiment with the service without the time commitment needed to oversee the printing and production. Bartellas added, for the service to be successful, there would need to be a framework that would support students and faculty throughout the process. Enabling students and faculty to come with an idea and be able to connect them with appropriate resources to get that idea off the ground would be what makes an effective service. Breaking down the service and not requiring ready-to-print models or the extra legwork of finding collaborators on campus will make the service successful; enable scholarship and research; and, as Bartellas said, allow the students to win.

Bartellas is an exemplary model of ingenuity in medical education and practice. He shows how medical students are engaging with technology and are using a maker mind-set to prepare for a future in innovative health care. With libraries stepping in and filling gaps in technology, they can enable more students like Bartellas and their faculty to achieve and accomplish even more in their education and careers.

CONCLUSION

Please let this chapter not only provide you with some insights outside the library but also show you the willingness and openness that you will find within the health innovator community. I was not expecting the depth of some of the responses I received, nor did I expect such quick responses.

After sending out one request for comment, I was shocked to see a reply within ten minutes. Not only did I receive a response, but I also received time to personally chat and discuss this topic further. One of the most prevalent things you will notice when you begin to venture into 3D printing and the maker movement is the collaborative nature everyone shares. Everyone wants to help each other succeed, and there is an excitement always lingering in every discussion and interaction. I strongly suggest, if you are starting a 3D printing service or looking to revitalize your 3D printing service, reach out to the innovators in your community and see what develops. You might find a partner or promoter for your service who can help engage your users and drum up interest and ongoing experimentation with the service.

REFERENCES

Indiana University. n.d. "Shannon Cooper, MSEd." http://iu.edu/~fmclerk/staff/cooper.html.

Indiana University–Purdue University Indianapolis. 2018a. "2016 CEG Awardees." Center for Teaching and Learning. https://ctl.iupui.edu/Programs/CEG/2016-CEG-Awardees/Scott-Renshaw-Shannon-Cooper-Zebulun-Wood.

———. 2018b. "Projects." CAMRI. http://www.iu.edu/~camri//Projects.php.

Interprofessional Practice and Education Center. 2018. "Leadership." Indiana University. June 20, 2018. https://ipe.iu.edu/about/leadership.

Lee, J. "Doctor as Designer: Speaking." http://www.doctorasdesigner.com.

Medicine + Design. 2018. "About—Rob." http://design-health.com/about.

Memorial University. 2018a. "Memorial University Faculty of Medicine Three Dimensional Printing." Biomedical Engineering. http://www.med.mun.ca/Biomedical-Engineering/MUN-MED-3D.aspx.

———. 2018b. "Ongoing Projects." Biomedical Engineering. http://www.med.mun.ca/Biomedical-Engineering/MUN-MED-3D/Ongoing-Projects.aspx.

Osmond, M. 2017. "A Lasting Difference: Student Recognized by Canadian Medical Hall of Fame." *Memorial University Gazette*. October 11, 2017. https://gazette.mun.ca/student-life/a-lasting-difference.

Perone, J., N. Anton, and A. K. Gardner. 2018. "The Various Roles of Simulation throughout the Surgeon Lifecycle." *Resources in Surgical Education*. https://www.facs.org/education/division-of-education/publications/rise/articles/lifecycle.

Purdue School of Engineering and Technology. 2017. "The Future of Cancer Treatment Is Now." April 19, 2017. http://www.engr.iupui.edu/main/about/news-events/news/2017/the-future-of-cancer-treatment-is-now.php.

Scott, C. 2018. "How to Start a Medical 3D Printing Lab." 3DPrint.com. https://3dprint.com/203749/medical-3d-printing-lab.

Stefanidis, D., N. E. Anton, L. D. Howley, E. A. Bean, A. M. Yurco, M. E. Pimentel, and C. K. Davis. 2017. "Effectiveness of a Comprehensive Mental Skills Curriculum in Enhancing Surgical Performance: Results of a Randomized Controlled Trial." *American Journal of Surgery* 213, no. 2: 318–24.

University of Michigan. 2018a. "Faculty Profile." School of Public Health. https://sph.umich.edu/faculty-profiles/lee-joyce.html.

———. 2018b. "Joyce Lee, M.D., M.P.H." http://ihpi.umich.edu/our-experts/joyclee.

University of South Dakota. 2018. "Introduction to Digital Design and Fabrication (3D Printing) Workshop for OT Practitioners." https://www.usd.edu/health-sciences/occupational-therapy/ot-workshops.

Zeek, A. 2017. "Indiana University CTSI Appoints New Deputy Innovation Officer." Indiana CTSI [Clinical and Translational Sciences Institute]. March 1, 2017. https://www.indianactsi.org/news_/new-deputy-innovation-officer.

Chapter Fourteen

Recommended Resources

This chapter presents resources to consult for additional information on the topics discussed throughout this book. Please note, this list is current as of July 2018. When things change, the information here should still be able to provide breadcrumbs to help find new content. Whether you are just getting started, wanting to learn more about printing, or interested in seeing what professionals are printing, this list can steer you in the right direction. Additionally, more in-depth information is provided through recommended books and journals.

BOOKS

For more in-depth information on 3D printing in medicine and 3D printing on its own, these books can support further learning. Some books focus specifically on 3D printing in medicine and others on learning 3D printing. For students and faculty interested in 3D printing, a tech shelf can also be made available with the printers or as a stand-alone section for browsing.

Aranda, Sean. 2017. *3D Printing Failures: How to Diagnose and Repair All 3D Printing Issues*. Edited by David Feeney. Amazon Digital Services. Get additional support for 3D printing problems. The book opens with practical advice for operating and managing 3D printers. The pictures included with diagnosing failures are very beneficial when starting out. The visuals help to better understand print failures and supply troubleshooting information to get printers back to normal. It can be difficult to troubleshoot online when you are new to printing and do not know the terminology; this book will help expand your knowledge and terminology to troubleshoot problems in the future.

Bell, Charles. 2014. *Maintaining and Troubleshooting Your 3D Printer*. Berkeley, CA: Apress. Prevent trouble before it starts by learning how to prepare for successful printing. This book offers background information on printing and guides you through calibration and software calibration. Information on maintaining printers is also featured to ensure machines perform as expected. Troubleshooting hardware and software failures is also included, as well as frequently encountered problems and solutions.

Bernier, Samuel N., Tatiana Reinhard, and Bertier Luyt. 2014. *Design for 3D Printing: Scanning, Creating, Editing, Remixing, and Making in Three Dimensions*. San Francisco: Maker Media. The basics of 3D printing are covered, along with designing models. Print file checklists are available to help prep models for printing. Advanced model capabilities are highlighted and include repairing and optimizing files for printing.

Chia, Helena N., et al. 2016. *3D Printing in Medicine*. Irvine, CA: Scientific Research. Dr. Helena Chia and her colleagues provide a more in-depth look at 3D printing in medicine. They review twelve publications to highlight current uses of 3D printing in health and medicine and discuss the future of 3D printing in medicine.

Gonzalez, Sara Russell, and Denise Beaubien Bennett. 2016. *3D Printing: A Practical Guide for Librarians*. Lanham, MD: Rowman & Littlefield. A book written by librarians, for librarians, *A Practical Guide* offers the nitty-gritty details on 3D printing. Policies, costs, training, marketing, and general printing guidance are also provided. This book explains how other libraries have established 3D printing services and supplements any areas where questions still linger.

Horne, Richard, and Kalani Kirk Hausman. 2017. *3D Printing for Dummies*. 2nd ed. Hoboken, NJ: John Wiley and Sons. Authors Horne and Hausman have backgrounds in engineering and IT and want to simplify 3D printing to make it accessible to everyone. This book provides the basics of 3D printing and offers instruction on creating your own models. Additionally, tips are available to help increase successful prints.

Kalaskar, Deepak, ed. 2017. *3D Printing in Medicine*. Duxford, UK: Woodhead. Edited by Dr. Deepak Kalaskar, lecturer in nanotechnology and biomedical engineering at the Division of Surgery and Interventional Science and the Centre for Nanotechnology and Regenerative Medicine, University College London (UCL). This book reviews the clinical applications of 3D printing, looking specifically at commercial and premarket uses. An overview of 3D printing, the different processes, and materials are covered. In-depth looks at medical applications of 3D-printed models, including patient-specific, surgical, pharmaceutical, and disease. Additionally, further advances into 4D printing are described.

Kelly, James F. 2014. *3D Modeling and Printing with Tinkercad: Create and Print Your Own 3D Models*. Indianapolis: Que. Learn the ins and outs of Tinkercad. This book breaks down the online 3D modeling software and explains the tools and features available to create custom models. These skills provide users with the ability to make their own models for prototyping. Additionally, some of the other skills beyond the technical details are discussed, including how to use 3D modeling and brainstorming ideas.

Kloski, Liza Wallach, and Nick Kloski. 2016. *Getting Started with 3D Printing: A Hands-on Guide to the Hardware, Software, and Services That Make the 3D Printing Ecosystem*. San Francisco: Maker Media. A new edition to be released in September 2018, this book provides instructions on the printing basics. CAD tutorials are also included. The book is a promising new addition to literature on developing 3D printing skills.

Rybicki, Frank, and Gerald T. Grant, eds. 2017. *3D Printing in Medicine: A Practical Guide for Medical Professionals*. New York: Springer Berlin Heidelberg. Dr. Frank Rybicki, also the editor-in-chief of the journal *3D Printing in Medicine*, together with Dr. Gerald T. Grant, composed a compilation of works by various medical professionals on the topic of 3D printing. The technology itself is explained with later chapters exploring the use of 3D printing in various specialties. Craniofacial, cardiovascular, musculoskeletal, and patient specific applications are discussed. Post processing of DICOM images and quality and safety issues are also included.

JOURNALS

There are only a few journals on the topic of 3D printing with applications in medicine and health care. Other health- and medicine-focused journals include articles on 3D printing, but if you're interested in adding a journal to your reading list, these titles might be useful.

3D Printing and Additive Manufacturing. https://home.liebertpub.com/publications/3d-printing-and-additive-manufacturing/621/overview. A broader journal that still focuses on 3D printing but in a variety of fields—medicine, education, food, and architecture. Also included in articles are discussions of new printing processes, tools, materials, and applications for use.

3D Printing in Medicine. https://www.springer.com/materials/journal/40861. An open-access, peer-reviewed journal that was started in 2015. The editor-in-chief is Dr. Frank Rybicki, professor and chair of radiology at the University of Ottawa and chief of medical imaging at the Ottawa Hospital. The journal highlights innovative applications of 3D printing in health and medicine and provides a platform for sharing STL files.

3D Research. https://www.springer.com/engineering/signals/journal/13319. A biomedical sciences journal that does not focus on 3D printing but instead focuses on the development of the 3D models. Topics include engineering, physics, materials sciences, chemistry, biology, and life and environmental science.

International Journal of Academic Makerspaces and Making. https://hemi-makers.org/ijamm. The goal of this journal is to share best practices among academic makers. This includes such topics as staffing levels, management, designs and layouts of spaces, training models, technology and equipment, safety and legal issues, maker activities, and how to determine impact of services.

Progress in Additive Manufacturing. https://www.springer.com/engineering/industrial+management/journal/40964. Academic, government, and industry are the main areas reviewed in this journal. Advancements in 3D printing are highlighted, along with data processing and simulation. Hybrid processes, the combination of additive and subtractive manufacturing, are also discussed.

MAGAZINES

Not only are these magazines good resources for you to review, but they are also great to put in whatever space features your 3D printers.

All3DP. https://all3dp.com. *All3DP* is an online-only publication that offers features and continuous postings highlighting new developments in 3D printing and sharing product reviews, project ideas, product deals, tutorials, and stories involving 3D printing. Tutorials range from 3D printer basics to advanced software modeling. Additionally, tutorials have a maker-esque element and include lessons on creating filament sensors and other combinations of 3D printing and electronics. There is no cost to view content, and *All3DP* is a great resource to review for tips and ideas.

Make:. https://makezine.com. Featuring stories, tested projects, tool guides, and maker spotlights, *Make:* magazine offers a wide breadth of information and resources to readers of all experience levels. The expert advice provided by the authors and commenters will help to fuel your library's 3D printing service. While not focused entirely on academic pursuits, *Make:* magazine provides the background knowledge of what's possible in making, and this information can be helpful in more advanced print requests. The website freely offers some material, and content and a print subscription is also available for $34.99 per year.

TCT. https://www.tctmagazine.com. Like the Rapid + TCT conference, the *TCT* magazine is geared more toward the industrial manufacturing crowd. It is a good resource to review the latest and greatest in manufacturing and gives an idea of what might be coming down the pike for consumer 3D printing. Information on 3D printing services, software, and scanning is available, as well as webinars for additional product information. The magazine is available freely online, and print editions are also available but may require payment depending on ability to meet qualifying criteria.

3D PRINTING COMMUNITIES

Finding a community within your organization is a great resource to have, but if you do not have this available to you or even if you want to increase your own community, these online communities can help troubleshoot and offer collaboration opportunities. While anyone can request to join these communities, group owners typically will ask questions in request forms about your interests of activities in 3D printing to verify that you are not there for other reasons (such as spamming).

3D Printer Tips, Tricks and Reviews. https://groups.google.com/forum/#!forum/3dprintertipstricksreviews. An active group that posts 3D printing tips, tricks, and project ideas. You can review posts without being a member, and once joined, you can make your own posts. A great resource for asking questions and reading about what others are doing with this technology.

3D Printing. https://www.facebook.com/groups/makerbot. Created more than six years ago, this Facebook group has more than 68,000 members. You must first request to join the group and answer a quick question about why you want to join. My request to join was approved in less than a day. On joining, you may first be overwhelmed with the amount of posts, but if you are new to printing, they are great for troubleshooting and getting excited about what can be made with 3D printers.

Thingiverse Group. https://www.thingiverse.com/groups/lib. Thingiverse Group is specifically for libraries and librarians. A forum provides an outlet for conversations, question and answer support, and the ability to share models using the Thingiverse platform. The unique element with this group is it is the only library- and librarian-specific group that I could find. Unfortunately, this group looks like it is either not active anymore or is not a busy group. It does offer a location for new members to join; activity may possibly be revitalized.

SOCIAL MEDIA

Social media is a great way to get quick snippets of information. News travels instantly on social media, and you can immediately discover the latest information related to 3D printing and making in health care. YouTube serves as a great resource for tutorials and troubleshooting printer issues. Online communities also provide a great amount of knowledge from the experienced users, and you can also learn from others' posts about problems they've encountered. I've followed several Twitter accounts to learn more about what is going on in the field and what professionals are doing.

YouTube Channels

Videos offer a quick way to get information on a topic. The ability to see what speakers are discussing, such as when troubleshooting printer problems, is extremely valuable and makes the process easier. Additionally, 3D printing is such a visual topic that seeing printers and models aids in the learning process.

Chuck Hellebuyck's Electronic Products (CHEP) Filament Friday. https://www.youtube.com/channel/UCsdc_0ZTXikARFEn2dRDJhg. CHEP provides users with tips on 3D printing using Tinkercad. With instruction on using Tinkercad, CHEP also provides instructions on prepping the models for printing and the setting used for successful prints.

Make:. https://www.youtube.com/user/makemagazine/videos. In addition to the magazine, *Make*: offers a YouTube channel with much of the same content as the magazine. Project demonstrations, news, product reviews, 3D printer tips, and troubleshooting are some of the topics included. Videos from maker faires are also available, as well as maker videos that highlight contributors' making abilities.

Make Anything. https://www.youtube.com/channel/UCVc6AHfGw9b2zOE_ZGfmsnw. Make Anything is a personal account of founder Devin Montes, a 3D printing enthusiast. Printer reviews, projects, printer and software tutorials, 3D printing tips and tricks, and maker videos are included.

Maker's Muse. https://www.youtube.com/user/TheMakersMuse. The mission of Maker's Muse is to provide reviews, tutorials, and guides. Three videos are produced each week. Video topics include tips and tricks, CAD for newbies, your 3D printing questions answered, Meshmixer tips, and budget projects.

MatterHackers. https://www.youtube.com/channel/UCDk3ScYL7OaeGbOPdDIqIlQ. MatterHackers offers users tutorials, product reviews, and practical advice for common printing problems. Project ideas are also provided and include a variety of ideas, one of which includes printing a synthetic cadaver.

Sanladerer, Thomas. https://www.youtube.com/user/ThomasSanladerer. Thomas Sanladerer provides guidance on 3D printing through tutorials and how-to videos. Printer reviews are also available, as are videos breaking down the printers themselves to better understand how they work. Printer materials are also covered, and videos are also included to improve printer safety.

Twitter

Twitter provides short, quick bites of information. You can follow a variety of users to get a good overview of a subject. You can learn about upcoming events, new publications, and see what others are up to in the world of 3D printing in medicine.

Embodi3D. https://twitter.com/Embodi3D. Keep connected with what is going on with Embodi3D and see the latest models that are available for printing, plus find out about new tutorials. Opportunities for collaboration are also promoted and can be shared with your library's faculty and students.

Health and Human Services (HHS) Idea Lab. https://twitter.com/HHSIDEALab. Follow tweets from the HHS Idea Lab. The HHS Idea Lab supports innovation and exploration that enhances and improves health care and services.

Itagaki, Dr. Mike. https://twitter.com/MikeItagaki. Learn more about Embodi3D as well as the creator behind it. In addition to promoting new models, printer tips and demos are included, as well as the latest information on innovations in health care.

JeffInnovation. https://twitter.com/JeffInnovation. Jefferson Innovation has an active Twitter feed showcasing the new and exciting ideas coming from their labs and information on current issues in health care.

Ku, Dr. Bon. https://twitter.com/BonKu. Dr. Ku also works with Dr. Pugliese (see chapter 13) in JeffInnovation. Follow Dr. Ku to learn more about design thinking in medicine and issues in public health.

Lee, Dr. Joyce. https://twitter.com/joyclee. Follow Dr. Lee to find more information on design thinking in medicine and to read about the latest information in public health.

Magnetta, Dr. Michael. https://twitter.com/MichaelMagnetta. Check out some of the 3D prints coming from the 3D Print Lab at the University of Pittsburgh Medical Center. Radiologist Dr. Michael Magnetta manages the print lab and showcases some of the finished models. Follow Dr. Magnetta's feed to see the possibilities for anatomy models.

MakerHealth. https://twitter.com/makerhealthco. See what is going on with maker health movements and what fellow makers have created.

MakerNurse. https://twitter.com/MakerNurse. Get inspired seeing what maker nurses are doing and get ideas for projects or events for your library. Find out about upcoming events and stay connected with MakerNurse's developments.

Mayo Innovation. https://twitter.com/MayoInnovation. Go beyond 3D printing and learn more about medical innovations and technologies with the Mayo Innovation account. Get some great ideas for events in which you can integrate 3D printing.

Morris, Dr. Jay. https://twitter.com/JayMorris_MD. Dr. Morris is a neuroradiologist from the Mayo Clinic and the codirector of the 3D printing lab. Check out Dr. Morris's Twitter feed for some awesome pictures of the latest prints coming out of the Mayo 3D print lab.

NIH 3D Print Exchange. https://twitter.com/NIH3Dprint. Find out the latest and greatest from the NIH 3D Print Exchange. See what other followers have printed, and learn about other relevant technologies, all within the same feed.

Pugliese, Dr. Robert. https://twitter.com/theEDpharmacist. Learn more about what is going on in JeffInnovation and the latest innovations in health care. Dr. Pugliese also posts about what the group at JeffInnovation has been doing in their community to promote public health.

UTMBMaker Health Space. https://twitter.com/UTMBMaker. See what a hospital maker health space gets up to, and read about new creations and events going on inside and out of the space.

3D MODEL REPOSITORIES

These repositories can act as ready references for finding models. The focus of the repositories should help you to determine which works best to find the model you need.

BodyParts3D. http://lifesciencedb.jp/bp3d. Models of the entire body are available, with detailed interior and external models. Models come from a sectioned-off DICOM data set and need to be joined together to create complete models. Furthermore, while DICOM models can sometimes be "messy" and need holes closed and smoothed, BodyParts3D offers cleaned-up DICOM models that should print without problems.

CG Trader. https://www.cgtrader.com. Offers a science models category that provides users the chance to upload and download models. Filter for free does not work well. Titles are not displayed on models themselves; however, when hovering over the model, you can view the text of the title. Also searches across different types of 3D models—animated, rigged, and more. If you can weed through the results, there are some gems. For instance, this is the first repository where I found a cleaned-up spleen with vessels and an inferior frontal gyrus. This repository includes premium and free content. Premium prices range from under five dollars to more than five hundred dollars.

Cults3D. https://cults3d.com. I had previously not heard of this repository, but I was impressed with what was available. Hovering over models reveals title and price. While it appears that a majority are premium models, there are free models available. This repository offers some unique anatomy and medical models.

Embodi3D. https://www.embodi3d.com. Embodi3D offers models created from DICOM data. Models are sorted into nine categories: bones, muscles, cardiac and vascular, organs of the body, skin, veterinary, science and research, miscellaneous, and medical CT scan files. Categories are further divided to make browsing easy and quick. Free files are available, but higher-quality models come with a cost.

GrabCad. https://grabcad.com. Offers a variety of less premium models than other broad repositories. Again, searching is somewhat tedious.

MyMiniFactory. https://www.myminifactory.com. Advertising models as guaranteed to print, models can be found by searching or browsing categories. The education category includes biology, chemistry, physics, mathematics, history, and design and technology.

NIH 3D Print Exchange. https://3dprint.nih.gov. Anatomy, biochemistry, and lab-ware models are freely available through this resource. Searching can be limited by category, and libraries have also been created to group together similar models. Print instructions are also available but should be reviewed.

Shapeways. https://www.shapeways.com. Traditionally thought of as an outsource service for 3D printing, Shapeways also offers a model repository made available through the shop. Some include downloadable files, while others require you to go through Shapeways itself for the printed model.

SPL Brain Atlas. https://github.com/lorensen/SPLBrainAtlas. Created by the Brigham and Women's Surgical Planning Lab Brain Atlas, students and faculty interested in neuroanatomy can benefit from the models found through this resource. The brain has been segmented into various parts and can be downloaded as STL files. Files can be searched using the "Find File" option.

stlFinder. https://www.stlfinder.com. Searches other 3D model repositories. Having previously searched the other repositories, I did find that it found models not listed in my earlier searches and repeated models that were found. It is a good tool to use but still requires searching the other repositories to avoid potentially missing anything.

Thingiverse. https://www.thingiverse.com. Thingiverse offers a wide variety of models, all available at no cost. Models can be sorted into groups or collections to help narrow to relevant models. Multiple variations of models are also available for some models. Users can use the comments to discuss model issues and offer a good resource for print directions based on user feedback.

Threeding. https://www.threeding.com. Offers models, but a majority are premium. This is a smaller repository that offers a good variety of anatomical models. One unique aspect of this repository is that it provides the price to download a model and a quote for printing. This is a nice alternative to consider if you are having printer problems or if the request is a higher quality or scale than you can print.

TurboSquid. https://www.turbosquid.com. Cost is much higher, but the quality is much better. The repository still includes a somewhat-random assortment of anatomy models, but the level of detail is higher than I have seen in other repositories (e.g., the structure of a normal heart artery).

Yeggi. http://www.yeggi.com. Searches other 3D model repositories. The ability to limit to free and paid models is possible, but no options are available to filter by category.

YouMagine. https://www.youmagine.com. While it doesn't have the quantities of medical and health models that other repositories have, this repository features models created by medical professionals.

SOFTWARE

These freely available software programs are for editing 3D models. The skill levels vary, and the software is listed in order from beginner to more advanced.

MakePrintable. https://makeprintable.com. This is a free service; however, it requires a phone number to send a verification code to before an account can be created. Models can be uploaded for a quick repair or advanced repair. The free account requires a "wait time" before your model can be processed. Side-by-side comparisons show the new versions of models and can be exported for printing. Model sizes must be under twenty-five megabytes to use.

Slicer. https://www.slicer.org. An open-source software, Slicer converts DICOM to STL and allows for patient-specific models to be printed. This software does require some time to learn to use.

MeshLab. http://www.meshlab.net. This is an open-source program that can do the basics—scale, orientation, and positioning—as well as advanced inspection and repair. The "healing" function can repair all sorts of model errors and help increase printability.

Meshmixer. http://www.meshmixer.com. Free 3D-model-editing software provided through Autodesk. This program offers editing and repairing options for your model and can help reduce print times by using the orientation and packaging features.

Netfabb. https://www.autodesk.com/products/netfabb/overview. Also, an Autodesk program, Netfabb is free to students and educators. Features include enhanced supports, optimized orientation, and editing to group parts together. Slicing and cutting is also available.

OnShape. https://www.onshape.com. A web-based editing software that can repair models and provide more advanced editing features. Free versions require models to be publicly available.

3D Slash. https://www.3dslash.net/index.php. Marketed as a "3D Piece of Cake," the main selling point of this software is its ease of use. A free plan is available for the online version of the software, which includes cloud storage. High resolution is not available in the free version; however, the premium plan is available for twenty-four dollars per year, and the educational plan includes a team management feature and no queue for export for ninety-six dollars per year.

Sculptris. http://pixologic.com/sculptris. Sculptris is a free program from Pixologic; the company also offers the more advanced program called Z Brush. Sculptris offers users an easy-to-use interface and is geared more toward designing, as users can create and edit models using sculpting tools that include pinch, inflate, flatten, smooth.

Adobe Photoshop. https://www.adobe.com/products/photoshop.html. Adobe Photoshop offers the ability to create 3D elements and can export files for 3D printing. This is not available freely and may require a higher level of skill to create 3D models.

SketchUp. https://www.sketchup.com. This is a free service that requires creation of a TrimbleID. Personal, educational, and professional use options are available.

Tinkercad. https://www.tinkercad.com. Another Autodesk program, Tinkercad is available online and offers numerous tutorials to learn how to create and design your own models. Aimed at K–12 educators and students, the program is easy to use and, being online, makes it a quick tool to access for basic editing. The ability to group models and create negative space in models is a unique feature that makes this a great go-to tool for 3D printing.

CONFERENCES AND EVENTS

If you are looking for professional development opportunities or just looking to connect with others in similar roles, these events are good opportunities to learn more about making and 3D printing.

3D Heals. An organization founded by neuroradiologist Dr. Jenny Chen, "3D Heals is a healthcare 3D printing innovation platform." The organization allows members across the world to connect and collaborate on 3D printing in health care. An annual conference is held, as well as other events throughout the year. Conferences include experts speaking on 3D printing technology and its use and applications in health and medicine.

Construct3D. A new conference, Construct3D is the result of collaborations between Ultimaker and Duke University. The first conference was held in 2017 and promotes academic use of 3D printing and digital fabrication for faculty, staff, and students. Professional development opportunities are also available with workshops and other learning activities throughout the event. Tours of local spaces are included.

FabLearn. An international conference that has been hosted in numerous countries around the world. The conference brings together teachers, professors, researchers, makers, designers, and policy makers. The goal of the conference is to provide learning opportunities to expand knowledge on digital fabrication and maker efforts.

International Symposium on Academic Makerspaces (iSAM). Part of the Higher Education Makerspaces Initiatives, iSam promotes collaboration between academic makerspaces. The conference includes workshops, activities, and demos of makerspace tools. Field trips to local makerspaces are also available, as well as networking opportunities.

Rapid + TCT. The Rapid + TCT conference is an event put on by the Society of Manufacturing Engineers and features many industry experts speaking about and demonstrating 3D printing technologies. The event is a broader conference, focusing on industrial uses of 3D printing, with smaller niche events discussing other uses. A medical manufacturing innovations track highlights 3D printing in medicine and new applications of the technology. The opportunity to speak with vendors is beneficial if you are interested in new equipment, and you can also identify places for outsourcing if requests become more advanced at your library.

DESIGN THINKING RESOURCES: FEATURING THE EXPERTS

Learn more about the experts who share their opinions in chapter 13. Find out more about their backgrounds and areas of interest and hear and read about the projects they are currently working on or have worked on in the past.

Doctor as Designer. http://www.doctorasdesigner.com. Dr. Joyce Lee provides a wealth of knowledge on design thinking and includes content based on her previous and current work on the subject. Also included are blog posts, recordings of speaking events, guides to accompany her workshops, and a variety of written works.

Innovation U. https://iu.mediaspace.kaltura.com/media/Pediatric+Grand+Rounds+1+17+2017A+"Innovation+U.+Patents%2C+Brainstorming%2C+and+Other+Keys+to+Thinking+Like+a+Biomedical+Innovator"++presented+by+Jonathan+Merrell%2C+MD/1_2ynhbais/64709041. Patents, brainstorming, and other keys to thinking like a biomedical innovator. Dr. Jonathan Merrell discusses the history of innovation in health and medicine and ties in his work with the Indiana CTSI and IU School of Medicine. He provides guidance in promoting innovation and highlights some student examples from groups he has worked with.

IU Health Technology Symposium. Hear from some of the experts on their ventures into innovation in medical education and health care. Shannon Cooper: https://tcs.iuhealth.org/tcs/#page:recordingList&pageNumber:1&id:309B330F-F14C-4D8D-B4B3-68BE5BFD7EBA; Dr. Robin Janson: https://tcs.iuhealth.org/tcs/#page:recordingList&pageNumber:1&id:77BBD15F-3215-4E69-883D-DE692B9317DB; Dr. Jonathan Merrell: https://tcs.iuhealth.org/tcs/#page:recordingList&pageNumber:1&id:E8405A8D-5959-451A-AA95-B0B82D438C28; Brian Overshiner: https://tcs.iuhealth.org/tcs/#page:recordingList&pageNumber:1&id:2755C9C0-DF12-4713-B0A7-739F380F41D6.

MISCELLANEOUS TOOLS AND RESOURCES

These resources support all the elements that go into 3D printing efforts, including printer operations, marketing, and data management.

3D Printing Coupons

One of the difficult things when starting 3D printing is knowing how models should look from a printer that is properly calibrated and performing normally. These coupons help to test your printer and troubleshoot any problems you may encounter. These are just a few examples of what is available to test your printers.

3D Benchy. https://www.thingiverse.com/thing:763622. Designed by Creative Tools, this is a fun little print that will test your printer's capabilities. The little boat model tests the precision of the printer and in the end is meant to float. Printing watertight requires the printer to be very precise, and any areas not up to par will cause either the print to fail or to not meet the standards meant for the model postprint (i.e., floating).

Make: 2017 3D Printer Test Files. https://www.thingiverse.com/thing:2755063. Put your printer to the test with these fourteen models. Created by *Make:*, these models not only test your printer but also help you to troubleshoot issues based on the outcome of the final prints. A link is available to assist in this troubleshooting, and I recommend giving some of these models a test to ensure that your printer is up to par.

Test Your 3D Printer! https://www.thingiverse.com/thing:1363023. Created by CrtlV, this model is more advanced and tests your printer's ability to print fine details. Through the model design, Z-gap, warping, rafts, and walls are all checked. In total, twenty-one tests are being run on your printer as it works on the model.

Marketing Tools

Marketing is challenging for libraries due to the multiple avenues and user groups to send information to. These tools offer some support in these efforts.

Buffer. https://buffer.com. To help manage multiple social media accounts, Buffer lets you schedule posts. This helps with planning out marketing campaigns and makes managing individual accounts a little easier.

Canva. http://www.canva.com. Need some help creating posts that are more visually engaging? Canva makes flyers, posters, and social media posts easier by offering templates, tons of freely available illustrations and photographs, and a plethora of tools to personalize content.

goo.gl. https://goo.gl. If you are posting links in your marketing campaigns, I recommend using goo.gl to track the hits from specific postings to determine the marketing efforts that are and are not working. You can track all links through a Google account. Tracked data include totals, referrers, browsers, countries, and platforms.

Hemingway Editor. http://www.hemingwayapp.com. For blogs or any other informational posts, it can be tricky keeping language from getting too technical or cumbersome. Hemingway Editor acts as the unbiased reader giving feedback about the content of posts and how to make them more readable.

Data Management Tools

These tools focus on supporting the development of data management plans. They are geared more toward researchers but can be tailored to fit a 3D printing service.

3D Printing Cost Management Excel Sheet. https://www.reddit.com/r/3Dprinting/comments/3e9icz/3d_printing_cost_management_excel_sheet. Created by a user on a Reddit 3D printing forum, this template can be used to begin the process of collecting data with the 3D printing service. Formulas and calculations are already set up to determine money spent and earned, filament usage, print time, and profit. It can be customized to fit individual services' needs. It is a great option if you want to get started quickly.

DMPOnline. https://dmponline.dcc.ac.uk. A tool to help you create a data management plan. This is an alternative if you want to try another DMP builder tool.

DMPTool. http://www.dmptool.org. Focused on assisting researchers meeting institutional and funder requirements, DMPTool allows for users to create data management plans that can be shared, downloaded, and stored on users' dashboard. Users can also select up to six different institutions and organizations for guidance when creating a data management plan. A default template is also provided for a quick start.

FDA Resources

While not necessarily tools, these resources are helpful to review for your 3D printing service.

"Technical Considerations for Additive Manufactured Medical Devices." https://www.fda.gov/downloads/MedicalDevices/DeviceRegulationandGuidance/GuidanceDocuments/UCM499809.pdf. An FDA guidance document. It might not be applicable to what you will be printing in your lab, but this guidance document can help prepare students and inform faculty who are interested in pursuing this area of 3D printing.

"Technical Considerations for Additive Manufactured Medical Devices." https://www.fda.gov/medicaldevices/newsevents/workshopsconferences/ucm587582.htm. Get a little more information on the FDA technical guidance with this webinar. Again, linking this through your 3D printing service page helps to inform your community about more advanced 3D printing in medicine.

Appendix A

Survey Results

1. Can you describe your library? (academic, hospital, medical, health sciences, etc.)

    - Academic health sciences
    - Hospital
    - Academic health sciences library
    - VA teaching hospital
    - Hospital
    - Health sciences library
    - Academic health sciences
    - Hospital
    - Academic
    - Academic health sciences
    - Academic, health sciences library
    - Academic health sciences
    - Academic health sciences
    - Health sciences university and hospital
    - I work at an academic health sciences library that is physically situated with the medical school and next door to the nursing school and hospital. This library is part of a large academic library system.
    - Academic medical

2. Does your library currently offer a 3D printing service?

    - Yes. (10; 62.5 percent)
    - No, but want to start. (4; 25 percent)

- No, still gathering information to decide whether or not to offer this service. (2; 12.5 percent)

3. What were/are the main obstacles to starting your service?

- Price and maintenance
- Money
- None. We are currently in the process of setting up a service. Cost recovery and initial investment are two of the larger considerations. Also having a space where limonene can be used.
- Have no idea what purpose it would serve in the library. Rehab has one for developing prosthetics already.
- Staffing . . . not knowing how to use the technology
- Money
- Money
- Cost and implementing policy
- Staffing, equipment selection and troubleshooting, material choices, IT support sources
- Interest. We knew we could get into 3D printing, but wanted to make sure that if we bought the stuff, people would use it.
- Getting users to try it or incorporating it into the curriculum
- Cost and skills training—a good printer is expensive, and building the skills to properly take advantage of it is very time-consuming.
- Cost, buy-in from faculty, staff training
- Ongoing funding support
- Learning about 3D-printing technology, designing an appropriate space, and providing professional development to a librarian and student workers
- Support from other departments on campus, support from administration/IT

4. What are some of your struggles with your service? (printer issues, software issues, staffing, etc.)

- Keeping up with production, the rate of jobs that are coming in are very high in comparison to how quickly the printer prints
- Justification
- Having to rely on IT person who is interested/knowledgeable about it (but who lacks time) because 1 FT staff member in library is busy
- Equipment issues
- Awareness and usage by students, residents, and faculty
- Logistics with our printer reservation system

- Desktop printers produce very basic prototype models. We don't have sophisticated equipment to make clinic-ready models for patient care environments. We don't have staff to help design/edit images.
- None so far. Just low usage.
- Our big printer has regular problems because we use it very heavily. When it stops and knocks over models, we have to use more material and restart projects, which can cause big unexpected delays and costs.
- Printer issues, finding funds for upgrading printers/computers
- N/a
- Printers break. Requests are overwhelming. Need more staff and space to continue growing.
- Not applicable

5. What do you charge for your service?

- 10 cents a gram
- N/a
- Right now we don't charge, but we plan to advertise and if needed charge for service. Also we have to write a policy.
- Seven cents per gram, I believe.
- $.10 per gram
- Free
- All services are free
- None so far
- We do not charge unless it is a special project that will use an unusual amount of material and staff time—for example, we are working on a project that involves printing large numbers of models over and over for one of our anatomists. He bought the materials and an extra print head. Our students and faculty can print one or two models per term at no charge because we have a special fund from our campus administration to cover costs for three years. They are enthusiastic about 3D printing.
- 0.10 cents per cubic gram and a $2.00 service fee. 3D scanning of an object is free. Class-related projects are currently free.
- N/a
- Free for academic, clinical, and research use. (We say we charge for personal use, but we don't get those requests ever.)
- Not applicable

6. Do you offer postprocessing (support removal and sanding, etc.) with your service?

   - Yes
   - N/a
   - No, haven't done that yet
   - No
   - Sometimes
   - Not formally, but we will if it's complex enough for us to worry about the user breaking their own piece.
   - Yes, but very basic. We may remove supports and advise on corrections that can be made to produce a more satisfactory product.
   - No
   - Yes, of course. Doing this requires equipment and skill that our students cannot reasonably be expected to have or develop.
   - Yes
   - N/a
   - Yes—support removal
   - Not applicable

7. What are some examples of what you have been requested to print? Are these requested by students, faculty, or staff members?

   - A faculty member requested to have a brain printed that he modified to look like one that had undergone stroke for students and patients to see.
   - Faculty members are interested in hearts and difficult airways. Students are interested in kidneys, pediatric skeleton, interior views of the skull, liver, adult skeleton, and bones of the hand and foot that are larger than scale.
   - N/a
   - A skeletal foot model from Podiatry. An eye model for Ophthalmology and dentistry is interested.
   - Anatomy models and orthopedics. These were requested by faculty and students.
   - Models for student projects; models for faculty.
   - Cosplay, microscope fittings, medical models, endless mechanical parts, pokemon, dnd figurines, prosthetics, sculptures, landscapes . . . everything you can think of.
   - Anatomical models. Lab equipment, branded giveaway magnets. Mobile device holder for microscopes. Scans of bones.
   - Models of body parts and one paperweight.

- Heart models in four slices, half-skulls, reproductions of a particular sculpture that is beloved by our campus, vertebrae (LOTS of them), hyoid bones and the connected neck, teeth, dental tools, assistive equipment, research pieces, lizard heads, etc. They've been requested by all three—we encourage it. The most unusual things are mostly faculty, but not all.
- Health related: Anatomy models (e.g., heart), OT devices. Misc.: school logo, Harry Potter glasses
- N/a
- Anatomical structures, machine parts, invention prototypes, etc., requested by students, faculty, staff, and clinicians
- Not applicable

8. Are you currently integrated in a curriculum or working with faculty in any other way?

- No
- We are currently exploring partnerships in this area.
- Resident orientation and research support
- No
- Yes
- Yes
- No
- Not integrated into curriculum at all
- No. We wish.
- Yes. Some members of the library staff worked with teams of students to apply for awards to do certain research projects based on 3D printing. We also provide models for specific courses to use.
- 1. Teaching in a first-year symposium week; 2. class project for OT students (students need to locate and print a OT device); 3. graded, two-week fourth-year elective.
- Yes, embedded into medical curriculum (not 3D printing yet)
- Faculty tell their students to visit the 3D printing lab at the library.
- Yes

9. Is there anything unique about your service or anything you'd like to mention not asked above?

- I have been exploring the possibility of using this technology in outreach.
- Both patient and medical library service—is this what your question meant???
- Not at this time

- No
- No
- We're fully staffed 12 hours a day, we couldn't make it work without trained student staff on hand to support. We use Prusa i3 MK2s and Hatchbox PLA filament exclusively.
- I wonder if a demo day would bring more attention to our service.
- I think that the most important two pieces for our success—and we do count ours a success—have been significant financial support from our administration and dedicated staff time. Because of those two pieces, we've been able to engage our students AND our faculty by having enough leeway for them to experiment, and have been able to give them the staff assistance they've needed to make their experiments a success.
- Currently exploring the role of 3D printing in data management.
- No
- Not applicable

Appendix B

3D Printing Data Collection: Fields

Use these fields for collecting data on 3D printing requests:

Request_Number
Requestor_Name
Requestor_Type
Requestor_Field
Requestor_Email
Date_of_Request
Request
Request_Intended_Use
File_Included
File
Request_Special_Instructions

Use these fields for collecting data on the print files:

Request_Number
Request_File_Name
Model_Weight
Model_Dimensions
Print_Time
Filament_Number
Model_Infill
Model_Resolution
Model_Build_Notes
Print_Success

Use these fields for collecting data on the filament usage:

Filament_Number

Material_Type
Material_Weight
Material_Open_Date

Glossary

accuracy. The correctness of information and content being published or promoted online.
acrylonitrile butadiene styrene (ABS). A popular material used with FDM printers. It is stronger than PLA material, but odor and fumes have been drawbacks that prevent more usage.
additive manufacturing. Another term used to describe 3D printing; typically, in industrial settings.
authority. The credibility and expertise of the author or owner of a resource that makes it better than others.
benchmarking. Tests that check the performance of the printer. Also referred to as "coupons."
binder jetting. A printing process that uses a binding agent to meld together layers of powder to create a solid shape.
blended/composition materials/filament. Filaments that are a combination of other materials. Examples include PLA with metals or wood.
build plate. Where the model is "built" or printed. Also known as the print bed.
build volume. The amount of space available where the model can be built. This determines the model size.
calibration/recalibrate. Tuning your printer to make sure all the settings are correct for best outcomes in printing. Calibration includes the Z-gap, build-plate leveling, and extruder.
computer-aided design (CAD). Creation of 2D or 3D models using computer software.
content marketing. Marketing specific content to specific audiences.
copyright. Allows for IP owners to profit from their work and prevents others from profiting or unfairly reusing or changing their work.
CRAAP. A method of evaluating a resource (currency, relevancy, authority, accuracy, purpose).
currency. The timeliness of a resource; is it recent enough to still be relevant, or is it outdated?
data. Statistics, numbers of any other value related to an object or other item.
data management. Ensuring that data is collected and stored appropriately to allow for it to be reproducible.
data management plan. A plan that determines the data collected, how it is collected, and where it is stored.
data probe. Gives the position of the mouse in the data set in DICOM software.
decimate. A process in editing a 3D model in which the number of polygons and triangles is reduced.
design thinking. Involves identifying a problem, applying knowledge learned to the problem, thinking of possible answers, developing a tangible solution, and testing and building on that solution.

digital imaging and communications in medicine (DICOM). Data sets that are produced by medical imaging machines; usually CT, MRI, and ultrasound.
extruder. The part of the printer that heats, melts, and pushes filament out onto the print bed, layer by layer, to build the model.
extruder temperature. The temperature the extruder heats to. Extruder temperature varies based on the type of filament.
fab lab. A variant of a makerspace that is only focused on 3D digital objects and not on printing.
fail friends. People to meet with and talk about your print fails.
FDA. Food and Drug Administration.
filament. The material used to produce a printed, layered model. This is typically ABS, PLA, or PETG.
fuse deposition modeling (FDM). A printing process that heats filament and then extrudes it through a nozzle onto a print bed, where it deposits it layer by layer until the final model is completed.
hackerspace. An early version of a makerspace that involves "hacking" objects to repurpose their original intended use.
Health Insurance Portability and Accountability Act (HIPAA). Protects patients from having their medical information released without their permission.
infill. How models are filled in as they print. Typically, lines, honeycomb, or diamond patterns.
information. Content derived from data.
intellectual property (IP). Ownership of creative or original works.
Kickstarter. Company that helps get early ventures off the ground with crowdsourced funding.
Kobayaski Maru. A fictional Star Trek training exercise in which there is no winning solution.
lab support models. Models used to help with lab work and may involve some prototyping.
layer height. The thickness of the layers.
liability. Holding someone responsible for damage or harm.
maker culture. The openness and "spirit of sharing" associated with making.
maker health. Using the maker mind-set to develop creative solutions to problems in health care.
maker mind-set. A creative or innovative thinking process that allows for people to make.
maker movement. The widespread efforts of people who make items to solve problems.
makerspace. A space dedicated to making and collaborating with others with dedicated equipment and supplies.
making. The act of creating something.
marketing. Advertising the value of services or products to users.
marketing cycle. The process of reviewing marketing.
marketing plan. The cycle of setting goals, researching target populations, identifying objectives, beginning marketing efforts, and then reviewing and modifying those efforts.
medical liability. When a health-care provider causes harm to a patient through a negligent act.
mesh. Ultimately the shape of a model based on the combination of the parts of the model—triangles, shells, edges, etc.
model weight. The weight of the material used to print a model.
modules. Functions set for manipulating the data within Slicer.
molecular and biochemistry models. Models used to better understand molecular biology and biochemistry.
nearly raw raster data (NRRD). The file formatting for image-processing applications.
noise. Data in a model that distracts the main model.
normal anatomy models. Models typically used as study aids and sometimes for "inspiration."
nozzle. The end of the extruder where the filament is pushed through to reach the build plate.
nylon. A plastic material used for its strength, durability, and lack of warping.
OBJ. A file type for 3D models. This type of model file contains more data about the digital version and can include texture and colors.
open-source. Software that is available freely and is not proprietary.
orientation. The position of a model on the build plate that includes the location and angle.

overhang. Parts of a model that do not have anything underneath them to support the material's weight. Depending on the angle, these most likely require supports to be added during the creation of the print file.
particulates. Fine particles of the filament that is ejected during printing that does not become part of the final model.
patents. Allows for the intellectual property owner to give permissions to others to reproduce their work, but this is only for inventions or new processes.
pathology models. Models used to learn more about abnormal anatomy.
polycarbonate (PC). A plastic material that is the strongest of the plastic family and can be transparent. Also known as the plastic used in eyeglasses.
polyethylene terephthalate glycol (PETG). A plastic, food-safe material that can be sterilized and is like a cross between PLA and ABS.
polylactic acid (PLA). A plastic material known to be the easiest material to print with and good when starting out with 3D printing. It is odorless when printing and the popular choice among academic print labs.
premarket approval. The FDA requirement for class III medical devices that have a more advanced intended use.
printable model/file. A model, usually STL or OBJ, that is designed to be 3D-printed with appropriate wall thicknesses and no cracks in the shell.
print bed. Where the model is built or printed. Also known as the build plate.
print coupons. Models used to test the performance and functions of the printer.
print file. The file that contains the sliced model and is specific to the printer.
print interview. Discussing the specifics and intentions of a user's requested model.
print speed. How fast the extruder moves across the print bed building the layers of the model.
print time. The total time to complete a model on the printer.
promoting. Advertising the availability of services or products to users.
protected health information (PHI). Information relating to a patient that could be used to identify them.
purge. Extruding filament without printing a model. This is usually done to clear a jam in the extruder or when changing material and checking the temperature with the new material type.
purpose. The intention of a resource. Is it to inform or advertise? Used when evaluating resources with the CRAAP method.
raft. Used when setting up a model for printing to help the model adhere to the build plate. It is also used to help prevent warping.
rapid prototyping. The original phrase to describe the production of 3D models for creating test products and manufacturing parts.
relevance. Relates to the content's intended audience when reviewing a resource.
replicating rapid prototyper (RepRap). The printer designed to be able to print its own replacement parts.
resin. Liquid material used in stereolithography printing.
resolution. The clarity of the final printed model. A higher resolution diminishes the visibility of layer lines and allows for finer details to be seen.
scale. The sizing of a model. A model might need to be scaled up or down to increase the size.
scene. When editing, the location of the data.
shells. The outer walls of the model.
social media rule of thirds. A marketing rule that divides social media postings into thirds. One-third of posts should be informative, one-third should be interactive with users, and the last third should promote.
stereolithography (SLA). A 3D printing process that uses ultraviolet light to cure a liquid material (resin) to build a model layer by layer.
stereolithography (STL). A file type for 3D models.
stress-reliever models. Nonmedical or noneducational prints that can serve as fun or are experimental.
structure. The parts of the anatomy.

subtractive manufacturing. Rather than material building up layer by layer to make a model from nothing, the model starts as a block of material, and layers are removed to create the model.

support density. The thickness of support material. This can affect the ability to break away and the amount of material used.

techshop. A space similar to a makerspace but requires a paid membership.

thermoplastic elastomers (TPE). A nonrigid plastic material that can create a flexible model.

thermoplastic polyurethane (TPU). A nonrigid plastic material that can create a flexible model.

Thingiverse. A 3D model repository offering a variety of free models for printing.

threshold. Upper and lower values that allow for certain densities of materials to be seen.

trademarks. Generally imply quality and that the standards of the product meet those set by the trademark owner.

transparency. Refers to the ability to see through the layers when editing a model.

wall thickness. Determines how well a model can print based on whether the walls can support the design. Walls that are too thin will not hold up during printing and create holes or cause your model to collapse.

warping. When the model curls away from the build plate or otherwise deforms as it cools too quickly after printing.

Z-gap. The space between the nozzle and the build plate. Too narrow, and your extruder will clog; too wide, and the material won't stick to the plate.

Index

About the Author

Jennifer Herron graduated from Ferris State University in 2010 with a BA in history and later graduated from Wayne State University in 2012 with her master's in library and information science. While completing her degree, she worked at a small health sciences college, providing library services to medical assistant students, among other health sciences students. She later moved on to work at a hospital library, where her interests in medical librarianship grew, as did her interests in supporting medical education.

Herron worked as the emerging technologies librarian for the Ruth Lilly Medical Library with Indiana University School of Medicine for more than three years. During this time, she helped to establish the library's 3D printing service and formed a university-wide 3D printing group that joined labs across multiple campuses throughout the state of Indiana. This collaboration between university 3D print labs ultimately led to a MakerHealth Faire, an interdisciplinary event bringing together students, faculty, and staff from medicine, information technology, engineering, health and rehabilitation, and informatics.

While establishing the library's 3D printing service, Herron collaborated with multiple departments to better understand the needs and wants for 3D printing from faculty, resident, and student perspectives. To fulfill requests, she learned the ins and outs of 3D printing through firsthand experience and from various online 3D printing organizations. Because of her knowledge and experience, Herron has advised other medical librarians on developing a 3D printing service and has presented locally, regionally, and nationally on 3D printing.

In her free time, Herron is creating her own personal 3D print lab and will continue to consult with librarians on 3D printing endeavors. She also hopes to partner with local public libraries to promote the use of 3D printing in patient education and health literacy.